SPACE SCIENCE, EXPLORATION AND POLICIES

FIELD PROPULSION PHYSICS AND INTERGALACTIC EXPLORATION

SPACE SCIENCE, EXPLORATION AND POLICIES

Additional books in this series can be found on Nova's website under the Series tab.

Additional e-books in this series can be found on Nova's website under the eBooks tab.

SPACE SCIENCE, EXPLORATION AND POLICIES

FIELD PROPULSION PHYSICS AND INTERGALACTIC EXPLORATION

YOSHINARI MINAMI
AND
HERMAN DAVID FRONING, JR.

Copyright © 2018 by Nova Science Publishers, Inc.

All rights reserved. No part of this book may be reproduced, stored in a retrieval system or transmitted in any form or by any means: electronic, electrostatic, magnetic, tape, mechanical photocopying, recording or otherwise without the written permission of the Publisher.

We have partnered with Copyright Clearance Center to make it easy for you to obtain permissions to reuse content from this publication. Simply navigate to this publication's page on Nova's website and locate the "Get Permission" button below the title description. This button is linked directly to the title's permission page on copyright.com. Alternatively, you can visit copyright.com and search by title, ISBN, or ISSN.

For further questions about using the service on copyright.com, please contact:
Copyright Clearance Center
Phone: +1-(978) 750-8400   Fax: +1-(978) 750-4470   E-mail: info@copyright.com.

NOTICE TO THE READER
The Publisher has taken reasonable care in the preparation of this book, but makes no expressed or implied warranty of any kind and assumes no responsibility for any errors or omissions. No liability is assumed for incidental or consequential damages in connection with or arising out of information contained in this book. The Publisher shall not be liable for any special, consequential, or exemplary damages resulting, in whole or in part, from the readers' use of, or reliance upon, this material. Any parts of this book based on government reports are so indicated and copyright is claimed for those parts to the extent applicable to compilations of such works.

Independent verification should be sought for any data, advice or recommendations contained in this book. In addition, no responsibility is assumed by the publisher for any injury and/or damage to persons or property arising from any methods, products, instructions, ideas or otherwise contained in this publication.

This publication is designed to provide accurate and authoritative information with regard to the subject matter covered herein. It is sold with the clear understanding that the Publisher is not engaged in rendering legal or any other professional services. If legal or any other expert assistance is required, the services of a competent person should be sought. FROM A DECLARATION OF PARTICIPANTS JOINTLY ADOPTED BY A COMMITTEE OF THE AMERICAN BAR ASSOCIATION AND A COMMITTEE OF PUBLISHERS.

Additional color graphics may be available in the e-book version of this book.

Library of Congress Cataloging-in-Publication Data

ISBN: 978-1-53612-554-2

Published by Nova Science Publishers, Inc. † New York

CONTENTS

Preface vii

Prologue xi

About the Authors xiii

Chapter 1 Introduction to Field Propulsion 1
Yoshinari Minami

Chapter 2 Energy for Spaceflight Power and Propulsion from Space Itself 37
Herman D. Froning Jr.

Chapter 3 The Evolution of Field Propulsion from Much Slower-Than-Light to Much Faster-than-Light Speed 59
Herman D. Froning Jr.

Chapter 4 Space Drive Propulsion: Typical Field Propulsion System 69
Yoshinari Minami

Chapter 5 Astrophysical Propulsion 123
Yoshinari Minami

Chapter 6	Galaxy Exploration: An Attempt to Begin It with an Initiative for Interstellar Flight *Yoshinari Minami*	**155**
Chapter 7	Rapid Transit by Field Propulsion to Distant Stars *Herman D. Froning Jr.*	**161**
Chapter 8	Hyper-Space Navigation *Yoshinari Minami*	**193**
Conclusion		**223**
Appendices		**225**
Index		**249**

PREFACE

Space development in the 21st century, unless there is a groundbreaking advance in the space transportation system, the area of activity of human beings will be restricted to the vicinity of the Earth forever and new knowledge cannot be obtained. The goal of traveling to the universe of humanity in the 21st century needs to extend not only to the manned solar system exploration but also to far-reaching stellar exploration in the future. This book addresses the latest advances in Space Propulsion Physics and Intergalactic Exploration.

Since the Wright Brothers first powered flight 115 years ago, power and propulsion for air and space travel has been by explosive combustion-heating and expansion of matter and expulsion of hot gases. And, for space travel, combustion-heating-expulsion takes place in the chambers of solid or liquid propellant chemical rockets for almost 90 years since Robert Goddard's first chemical liquid rocket launch in 1928. But now, almost 90 years later, chemical rockets have reached a plateau in the thrusting performance that they can achieve. And though remarkable, it is woefully insufficient for propelling spaceships economically, swiftly, safely over the enormously long gulfs of inter-planetary and interstellar space that much of humanity dreams of going. Also, the consuming-heating and emitting of gases, which has met Earth's terrestrial power and propulsion needs for over 150 years is now heating its atmosphere at unacceptable rate. So the authors believe now is the time to begin serious exploration of a nearly propellant-less mode of power and propulsion - one that creates energy, power, thrust by actions and reactions of mass-less fields – not by combustion and emission of mass.

The authors have been exploring this new mode of impulsion and power for many decades, while engaging in aerospace industry for aircraft development and satellite development, and they have reported on field propulsion research in peer-reviewed journals and conferences around the world. This book gathers together much of this work.

Thus, Chapter 1 presents basic principles of field propulsion; critical influences of the physical background where they move; and key features of the type of impulsion they develop. Chapter 2 introduces the physics of the zero-point quantum vacuum of space and possibilities and problems of interacting with this vacuum for field propulsion and power. Chapter 3 gives a perspective on how field propulsion might logically begin by augmenting existing types of propulsion systems; and then exploit other emerging technologies to evolve into ever more effective field propulsion systems which enable faster field-propelled systems that consume less and less propellant while achieving the greater and greater flight safety needed to travel over the ever longer distances needed to reach ever more distant planets more distant stars. Chapter 4 shows space drive propulsion as a typical example of field propulsion; with respect to mechanical structure of space, propulsion theory, propulsion mechanism. Furthermore space drive propulsion method from the aspect of cosmology is explained. Chapter 5 shows how astrophysical features of our universe hint at attractive features for field propulsion system that would propel craft through it. These features are shown in detail in an awarded patent for a field propulsion system. Chapter 6 combines the difficult problem of field propulsion with the equally difficult problem of interstellar navigation to fulfil the seemingly impossible dream of intergalactic exploration. Chapter 7 introduces a method – a possibility - for overcoming the "light barrier" (the seeming "wall-of-light" in 4-D space-time) that would prevent faster-than-light flight relative to Earth. And this would be done by "jumping" over this barrier in a higher-dimensional "hyperspace". And Chapter 8 introduces a hyperspace navigation theory to "jump the light-barrier" in the higher hyperspace.

In this respect, many excellent books with similar titles and similar books have already been published on aspects of field propulsion, space-vehicles and interplanetary and interstellar flight. But we believe this book is unique in giving readers a broad, yet in-depth view of field propulsion science and technology- one which includes astronomy, astrophysics, and interstellar navigation and advanced vehicle engineering.

Finally, the book concludes that spaceflight must ultimately be by actions and reactions of fields, not combustion and expulsion of mass. Although it emphasizes the extraordinary potential of field propulsion, this propulsion will only be achieved by significant advancement of other technologies as well.

Though the field propulsion system examples shown in this book will surely be those systems that are finally perfected by future scientists and engineers. And these people will have had to solve new field propulsion problems not known about right now. Nevertheless the authors hope that a few of these future field propulsion scientists and engineers may have been readers of this book.

Yoshinari Minami, Herman D. Froning, Jr

Prologue

For more than one hundred years since the world's first powered flight by the Wright Brothers in 1903, power and propulsion for air and space travel has been the explosive combustion and heating of matter in a chamber and expulsion of its hot gases. But this may eventually end; such consuming, heating, and emitting of matter cannot propel spaceships swiftly enough or far enough to go where much of humanity wants to go. This consuming, heating, and emitting of gases, which has met Earth's terrestrial power and propulsion needs for an even longer time, is now heating Earth's atmosphere at unacceptable rate. So we believe this is a good time to write about a potentially, nearly propellant-less mode of power and propulsion that creates energy-power-thrust by actions and reactions of mass-less fields – not combustion and emission of matter.

To give readers the broadest possible view of field power and field propulsion, the authors have combined their different talents. Thus, Minami (who did most of the work in preparing and editing this book) has set forth the mathematical and science foundations of field propulsion, as applied to both slower- and faster-than-light flight, while Froning has described some requirements and considerations for embodying field power and propulsion in flight vehicle systems. While Mr. Minami's presentations are more scholarly in approach, Mr. Froning's are more in the form of his personal experience in analysis, simulation and design of jet and field-propelled craft.

ABOUT THE AUTHORS

HERMAN DAVID FRONING, JR.

Herman D. Froning Jr. worked at the leading-edge of aeronautical and spaceflight science and technology for over 50 years, including much of the exciting 1945-1985 time-period when the greatest increase in air and space flight progress above Earth occurred. He performed research and development work for the U.S. Air Force, Boeing, McDonnell Douglas, and his own Company, Flight Unlimited. Examples of his pioneering work include design of the configuration concepts of the United States first anti-ballistic missiles and he was among the first to explore use of zero-point fluctuation energies of space for space power and propulsion. His own company, Flight Unlimited, performed work for the U.S. Air Force; National Aeronautics and Space Administration; European Space Agency; ANSER; Allegany Ballistics Laboratory; Thiokol Corporation. He is a Past Fellow of the British Interplanetary Society; Associate Fellow of American Institute of Aeronautics and Astronautics; and Past Member of the International Academy of Astronautics. He has served on various technical panels and was a Founding Participant in the NASA "Breakthrough Propulsion Physics" Program.

YOSHINARI MINAMI

Yoshinari Minami received his B.S. degree in electrical engineering from Ritsumeican University in 1971. He joined NEC Corporation in 1971. He has been engaged in the design and development of telemetry, tracking, and control (TT&C) subsystem and Data Handling System of many Japanese satellites in the Space Development Division.

After that, he has been engaged in the design and development of Japanese Experimental Module (JEM) in the Space Station Systems Division. Since then, he has been engaged in the patent control and intellectual property right in Network Engineering Department IA Creation Services Division.

He is now *administrative director* and project manager of *Advanced Space Propulsion Research Project* in Advanced Science-Technology Research Organization. He is a member of the Institute of Japan Society for Aeronautical and Space Sciences, and also a past member of the Institute of the Physical Society of Japan. Furthermore, at present, he is a member of IAA (International Academy of Astronautics) Scientific Committee, a past member of NASA BPP (Breakthrough Propulsion Physics) Group, and a past Fellow of the British Interplanetary Society.

Chapter 1

INTRODUCTION TO FIELD PROPULSION

Yoshinari Minami

1.1. PROLOGUE

The breakthrough of propulsion method has been required until now for the purpose of interplanetary travel and interstellar travel. Instead of conventional chemical propulsion systems, field propulsion systems, which are based on the General Relativity Theory, the Quantum Field Theory and other exotic theories, have been proposed by many researchers to overcome the speed limit of the conventional space propulsion system. Field propulsion system is the concept of propulsion theory of spaceship not based on usual momentum thrust but based on pressure thrust derived from an interaction of the spaceship with external fields. Field propulsion system is propelled without mass expulsion. The propulsive force as a pressure thrust arises from the interaction of space-time around the spaceship and the spaceship itself; the spaceship is propelled against space-time structure.

The field propulsion principle is based on the assumption that space as a vacuum possesses a substantial physical structure. Minami proposed a hypothesis about mechanical property of space-time in 1988 [1]. The substantial physical structure of space-time is based on this hypothesis. Field propulsion is propelled receiving the propulsive force (i.e., thrust) arises from the interaction of the substantial physical structure. The meaning of substantial physical structure regarding the space-time can be conjectured from both General Relativity in the view of macroscopic structure and Quantum Field Theory in the view of microscopic structure. Therefore, several kinds of field propulsion can be proposed by making choice of physical concepts. However, even if any propulsion system is selected; the propulsion principle of field propulsion system is the identical in regard to utilize the substantial physical structure of space.

As shown in Figure 1, the propulsion principle of field propulsion system is not momentum thrust but pressure thrust induced by a potential gradient or pressure gradient arising from the space-time field (or vacuum field) between the bow and the stern of spaceship. Since the pressure of the vacuum field in the rear vicinity of spaceship is high, the spaceship is pushed through the vacuum field. Pressure of the vacuum field in the front vicinity of the spaceship is low, so the spaceship is pulled by the vacuum field. In the front vicinity of spaceship, the pressure of the vacuum field is not necessarily low but the ordinary vacuum field, that is, just only a high pressure of vacuum field in the rear vicinity of the spaceship.

Propulsion Principle

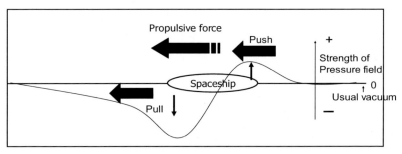

Asymmetrically interaction with the pressure of field creates propulsive force for the spaceship.

The strength of pressure field ahead of the spaceship is diminished and its behind increased, this would result in favorable pressure gradients.

Figure 1. Fundamental propulsion principle of Field propulsion.

1.2. DEFINITION OF FIELD PROPULSION

The term "Field Propulsion" and its theoretical concept were first defined by Minami. He introduced this concept in his publication [2] and strove to have it gain broader acceptance in this field of study. The detailed concept is summarized in [Y. Minami, "An Introduction to Concepts of Field Propulsion," *JBIS*, 56, pp.350-359 (2003)].

He defined the expression as follows: *Field propulsion is the propulsion method using the field, which is propelled by being pushed or pulled by surrounding space-time as a field. Field propulsion is propelled receiving the propulsive force (i.e., thrust) arises from the interaction of space-time around the spaceship and the spaceship itself, that is, the field propulsion is propelled against space-time structure. The field propulsion principle is based on the assumption that space as a vacuum possesses a substantial physical structure. The field propulsion utilizes the mechanical property of space-time possessing structure (continuum mechanics in the view of macroscopic structure, statistical mechanics in the view of microscopic structure). However, a propulsion using pressure thrust induced by mutual interaction with an independent matter, such as plasma, ion, and photon with unrelated to the structure of space-time is not applicable to field propulsion. The pressure thrust in field propulsion is different from the pressure thrust in solar sails or light sails, which is induced by receiving pressure of light.*

As described above, field propulsion is the concept of propulsion theory of spaceship not based on usual momentum thrust but based on pressure thrust derived from an interaction of the spaceship with external fields. The field propulsion principle is based on the assumption that space as a vacuum possesses a substantial physical structure. Field propulsion is propelled against space-time structure receiving the propulsive force (i.e., thrust) arises from the interaction of the substantial physical structure.

Instead of conventional chemical propulsion systems, field propulsion is based on the General Relativity Theory, Quantum Field Theory and the other exotic theory. In this chapter, the basic concepts for such means of propulsion within the present physics paradigm are presented.

1.3. Physical Structure of Space as a Vacuum

As has been previously mentioned, the propulsion principle of field propulsion is based on the assumption that space as a vacuum possesses a substantial physical structure. Field propulsion is propelled receiving the propulsive force (i.e., thrust) arises from the interaction of the substantial physical structure. Then, what is the meaning of substantial physical structure regarding the space-time? The answer can be conjectured from both General Relativity in the view of macroscopic structure and Quantum Field Theory in the view of microscopic structure. Firstly, General Relativity is the geometric theory of gravitation and the gravitation is explained by curved space. The curvature of space plays an important role. Although the curvature is quantity in mathematics, the curvature has relations with continuum mechanics such as expansion, contraction, elongation, torsion and bending. This physical relation indicates that the space as a vacuum can be considered as a kind of elastic body in continuum mechanics. Therefore, the propulsion system used General Relativity is to be proposed from the standpoint of continuum mechanics. Secondly, space-time as a vacuum is generally viewed as a transparent and ubiquitous infinitive empty continuum, upon which physical events take place. However, Quantum Field Theory and quantum electrodynamics (QED) view it as possessing vigor and vitality over scales of time and space. Such vigor and vitality are the zero-point fluctuations of the vacuum electromagnetic field (vacuum perturbation), and the continuous creation and annihilation of virtual particle pairs are iterated there. Further, according to the latest quantum optics, until recently, it has been considered that the control of vacuum perturbation was utterly impossible. At the present day, it is proven that the vacuum perturbation can be controlled by squeezed light technology. It is possible to increase the energy density locally above the vacuum state and, vice versa, decrease the energy density locally below the vacuum state. That is, the squeezed light generates the squeezed vacuum states and yields the coordination geometry of energy density.

Furthermore, the strings of superstring theory are considered as the threads of the space-time fabric. String seems to be the fundamental element of the substructure or fine structure of space-time. Supposing that the string is the constituent of space-time, it is suggestive of the existence of possible quantum states for space-time. This indicates that entropy of space-time can be defined as an assembling of strings. Strings as the constituents of space-time correspond to the polymer chains in the elastic body. Since the statistical entropy is the logarithm of the number of states (i.e., degeneracy of system), it is necessary to consider what kinds of physical state exist. Therefore, the propulsion system used Quantum Field Theory is to be proposed from the standpoint of quantum physics.

Thus, although several kinds of field propulsion can be proposed by making choice of physical concepts, the propulsion principle of field propulsion system is the identical in regard to utilize the substantial physical structure of space, even if whether the constituents of physical structure are curvature, zero-point fluctuations, or statistical entropy of string and so on.

1.4. Technical Status and Problem of Present Propulsion System

As described above, all kinds of current propulsion system except solar sail and light sail are based on the momentum conservation law. In the case of the momentum thrust based on momentum conservation law, the maximum speed (V) is limited by the product of the gas effective exhaust speed (w) and the natural logarithm of mass ratio (R). Here, Isp is specific impulse(s).

$$V = w \cdot \ln R = gI_{SP} \cdot \ln R. \tag{1}$$

The maximum speed V which a rocket can reach is theoretically determined by the gas jet speed w (m/s) and the mass ratio R.

Because the velocity of present rocket is too slow as compared with the speed of planet, the interplanetary exploration by mankind, not to speak of interstellar exploration, has various technical difficulties. We need the super-high speed and high acceleration of spaceship.

For example, the origin of the problem that the manned Mars exploration takes long-term time is due to the cruise velocity of a spacecraft being too slow. The second astronomical speed (11.2 km/s) that a rocket obtains for earth escape is slightly slow compared with the orbital speed (24 km/s) of Mars, and the orbital speed (30 km/s) of the Earth.

This is because the maximum speed of a rocket is limited by the product of gas effective exhaust speed and the natural logarithm of mass ratio (about its value is 7). The speed beyond this cannot be theoretically taken out from the propulsion principle of a rocket based on the momentum conservation law.

Concerning a chemical rocket which has multi-stage composition, about 10km/s speed is a practical limit. In the case of chemical rocket, its specific impulse I_{SP} is 460 seconds, so the maximum speed becomes 4.5km/s for single stage rocket. If the speed of a rocket is 1000 times quick compared with the speed of Mars or the Earth, a straight line orbit can be attained. Whenever you like always, it can reach to the target planet in a short time without restriction of orbital calculation, a start time, and return time, so that it may operate by car as it were.

Equation (1) can be represented as follows in detail:

$$V_f - V_i = \Delta V = \int_0^T \alpha dt = \int_0^T \frac{F}{m} dt = \int_0^T \frac{I_{SP}(-\dot{m}g)}{m} dt = I_{SP} g \ln \frac{m_i}{m_f} \quad (2)$$

Eq. (2) indicates that the speed increment ΔV of rocket when the rocket of the initial mass m_i reduces mass to the rocket of the final mass m_f by combustion for T seconds. Since the propellant mass m_p is given by Eq. (3), combining Eqs. (2) and (3) yields the Eq. (4).

$$m_p = m_i - m_f, \quad (3)$$

$$m_p = m_i\left[1 - \exp\left(-\frac{\Delta V}{gI_{SP}}\right)\right] \tag{4}$$

By expelling the mass of a propellant m_p outside, a rocket obtains thrust and increases the speed of ΔV, that is, the propellant is indispensable for the conventional propulsion system based on momentum thrust. Further, since a large thrust is required for the large weight of payload, a large amount of propellant is needed for rocket; therefore the increased weight of propellant, i.e., the rocket needs larger amount of propellant.

Accordingly, we need the new propulsion principle to exceed the limits of prior propulsion technology and seek entirely different technology.

As described in NASA BPP (Breakthrough Propulsion Physics), we must investigate the new propulsion system based on physics as follows.

1. *Mass:* Discover new propulsion methods that eliminate the need for propellant or beamed energy.
2. *Speed:* Discover how to circumvent existing limits (light-speed) to dramatically reduce transit times.
3. *Energy:* Discover new energy methods to power these propulsion systems.

1.5. FEATURES OF FIELD PROPULSION

A promising field propulsion system which gets ahead of conventional propulsion systems (i.e., chemical propulsion, electric propulsion, laser propulsion, nuclear propulsion) utilizes the ubiquitous infinite space, more specifically, vacuum. An extraction of thrust from the excited quantum vacuum is indispensable to developing field propulsion. Field propulsion is the prerequisite technological developments for mankind to expect in the 21st century. Because, mankind cannot realize the space travel by conventional propulsion principle based on expulsion of a mass to induce a reaction thrust (momentum thrust). As is well known, in momentum thrust propulsion, the maximum speed is limited by the product of the ejected gas effective exhaust velocity and the natural logarithm of the mass ratio; so the speed of present rockets is too slow as compared with the speed of planets. Alternatively, field propulsion has made advances resulting from pressure thrust, without mass expulsion. The propulsive force as a pressure thrust arises from the interaction of space-time around the spaceship and the spaceship itself, the latter being propelled against the space-time structure [1, 2, 3, 4, 5]. Although several kinds of field propulsion system are proposed, there exists one common technology for all kinds of field propulsion, which strongly depends on excitation of the vacuum.

On the other hand, there are attempts to explain gravitation and inertia from the viewpoint of quantum vacuum as follows. Gravitation and inertia are an induced effect brought about by changes in the quantum fluctuation energy of the vacuum when matter is present. Gravitation and inertia of matter originate in electromagnetic interactions between the zero-point field (ZPF) and the quarks and electrons constituting atoms. Especially, the attractive gravitational force is akin to the induced van der Waals and Casimir forces [6, 7]. Although the aforementioned theory is neither universally accepted nor proved at present, it is worthwhile recommendation. These play a successful role regarding the fine structure of vacuum. At the present time, although the empirical Casimir forces cannot be directly applied to propulsion, Casimir effects is the sole theory and experiment which indicates the possibility of extracting thrust from vacuum. In order to extract the micro-thrust from vacuum as a first stage, the perturbation of vacuum and the generation of localized inhomogeneous field are indispensable. Here, we define the perturbation of vacuum as the inhomogeneous field of the energy density locally below or above its value in the vacuum state. The vacuum fluctuates with oscillations of the electric field, and these vacuum fluctuations and the zero-point energy have a common origin in the quantized vacuum. The vacuum fluctuations are considered as quantum noise.

Until recently, it has been considered that the control of vacuum perturbation was utterly impossible. However, at present, it is proven that the vacuum perturbation can be controlled by squeezed light technology. Therefore, it is possible to increase the energy density locally above the vacuum state and vice versa, decrease the energy density locally below the vacuum state. That is, the squeezed light generates the squeezed vacuum states and yields the coordination geometry of energy density. The theoretical possibility of extracting thrust from the excited vacuum (i.e., squeezed vacuum state) induced by the control of squeezed light and the experimental concepts are also described here.

The field propulsion system appears to violate the conservation law of momentum because the reaction mass is not readily apparent. NASA considers that conservation of momentum can be satisfied in various ways that do not require having an on-board supply of reaction mass about conservation of momentum, as follows [8]:

1) conservation by using the contents of space (Interstellar Matter, Star Light, Magnetic Field, Cosmic Microwave etc.) as the reaction mass, 2) conservation by expelling non-mass momentum (equivalent momentum $P = E/c$) such as hypothetical "Space Waves", 3) conservation by negative mass, 4) conservation by coupling to distant masses via the intervening space.

However, the most promising interpretation is to consider that space itself as vacuum is a kind of reaction mass.

Figure 2 shows fundamental propulsion principle of Field Propulsion. As shown in Figure 2, the propulsion principle of field propulsion system is not momentum thrust but pressure thrust induced by a potential gradient or pressure gradient arising from the space-time field (or vacuum field) between the bow and the stern of starship. Since the pressure of the vacuum field in the rear vicinity of starship is high, the starship is pushed through the vacuum field. Pressure of the vacuum field in the front vicinity of the starship is low, so the starship is pulled by the vacuum field. In the front vicinity of starship, the pressure of the vacuum field is not necessarily low but the ordinary vacuum field, that is, just only a high pressure of vacuum field in the rear vicinity of the starship. The pressure gradient of the vacuum field (potential gradient) is formed over the entire range of the starship, so that the starship is propelled by being pushed from the pressure gradient of the vacuum field. To make the starship independent of the pressure gradient of the vacuum field, this propulsion system is essentially defined as a pulse propulsion system. In general, a body cannot move carrying, or together with, a field that is generated by its body. In other words, the body cannot move unless the body is independent of the field. Therefore, we must keep in mind that propulsion principle is not a kind of "surfing".

The most remarkable features attainable through Field Propulsion are as follows: 1) high acceleration such as several ten G can be obtained, 2) theoretical final velocity close to the speed of light, 3) no action of inertial force.

As to item 3), this comes from the thrust as a body force. Since the body force they produce acts uniformly on every atom inside the starship, accelerations of any magnitude can be produced with no strain on the crew, i.e., it is equivalent to free-fall. Therefore, the flight patterns such as quick start from stationary state to all directions in the atmosphere, quick stop, perpendicular turn, and zigzag turn are possible.

For instance, as an example of field propulsion system, the Space Strain Propulsion System [1] and alternatively the Space Drive Propulsion System [3, 4] were introduced by Y. Minami. The expression of "space strain" was changed to "space drive" after recommendation by R.L. Forward. Space Coupling Propulsion System was introduced by Marc G. Millis [8].

At present, physics admits the Zero Point Energy (Media of Electromagnetic Fluctuations of the Vacuum: Zero Point Fluctuations or Zero Point Field). It is said that the Zero Point Field is related to both gravitation and inertia. Space as a vacuum is a kind of actual field, which repeats the creation and annihilation of particle and anti-particle continuously. ZPF (Zero Point Field) propulsion system was introduced by H.D. Froning and T.W. Barrett [9, 10]. The various standpoints for engineering the Zero-Point Field were introduced by H.E. Puthoff, et al. [11].

The constituents of pressure gradient or the potential gradient generated by the propulsion system described above are as follows: curvature, entropy, zero-point radiation pressure, metric. The distribution of field as shown in Figure 2 is fundamental; accordingly, several kinds of propulsion systems have been proposed. Any propulsion system is selected, whether the constituents of pressure gradient or potential gradient generated by propulsion engine are curvature, metric, zero-point radiation pressure or entropy, the propulsion principle is the same principle as shown in Figure 2.

STARSHIP: Flight Principle

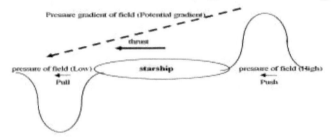

Zero-Point Radiation Pressure is KEY Factor
Potential Gradient is generated by Gradient of Vacuum Energy Density
Vacuum Energy Density is generated by Squeezed Light

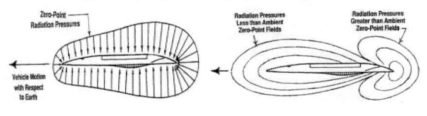

By Dave Froning

Figure 2. Fundamental Propulsion Principle of Field Propulsion.

Also we must keep in mind the following: "There is no propulsion theory exceeds the speed of light", moreover, in the real physical flat space, special relativity works correctly, and no propulsion theory will exceed the speed of light. Even with field propulsion, the maximum speed of field propulsion is theoretically quasi-light speed.

Now, a field propulsion system must satisfy the following criteria [2, 12]:

1) conservation of momentum, 2) conservation of energy, 3) ability to induce a unidirectional acceleration of the spaceship, 4) controllability of direction and thrust, 5) sustainability during spaceship motion, 6) effective capability of propelling the spaceship.

Referring to the figure on the lower side of Figure 2, the radiation pressures are less than ambient zero-point fields in the bow of starship, on the while, the radiation pressures are greater than ambient zero-point fields in the stern of starship. The pressure gradient of the vacuum field is formed over the entire range of starship, so that starship is propelled by being pushed from the pressure gradient of the vacuum field induced by Zero-Point Radiation Pressure [10].

1.6. Propulsion Principle of Field Propulsion

All existing methods of propulsion systems, i.e., chemical propulsion, electric propulsion (Ion thruster, MPD [Magneto Plasma Dynamic] thruster, Hall thruster, ARC jet thruster), laser propulsion, nuclear propulsion are based on expulsion of a mass to induce a reaction thrust. The "momentum thrust" is based on momentum conservation law.

Alternatively, the concept of "field propulsion" has been advanced as resulting from pressure thrust, without mass expulsion. The envisaged solar sails and light sails are propelled just by receiving light pressure, but pressure thrust in field propulsion refers to a reaction with space-time itself (i.e., the vacuum) to generate a propulsive force. The propulsive force as a pressure thrust arises from the interaction of space-time around the spaceship and the spaceship itself, the latter being propelled against space-time structure. The field propulsion principle consists in the exploitation of the action of the medium field induced by such interaction and is thus based on some concepts in modern physics to be found in General Relativity, Quantum Field Theory, Quantum Cosmology and Superstring Theory including D-brane in order to bring about the best propulsive performance.

Generic Propulsion Principle of Field Propulsion

As shown in Figure.3, the propulsion principle of field propulsion system is not momentum thrust but pressure thrust induced by pressure gradient (or potential gradient) of space-time field (or vacuum field) between bow and stern of spaceship. Since the pressure of vacuum field in the rear vicinity of spaceship is high, spaceship is pushed from vacuum field. Pressure of vacuum field in the front vicinity of spaceship is low, so spaceship is pulled from vacuum field. In the front vicinity of spaceship, the pressure of vacuum field is not necessarily low but the ordinary vacuum field, that is, just only a high pressure of vacuum field in the rear vicinity of spaceship. The spaceship is propelled by this distribution of pressure of vacuum field. Vice versa, it is the same principle that the pressure of vacuum field in the front vicinity of spaceship is just only low and the pressure of vacuum field in the rear vicinity of spaceship is ordinary. In any case, the pressure gradient of vacuum field (potential gradient) is formed over the entire range of spaceship, so that the spaceship is propelled by pushing from the pressure gradient of vacuum field.

Here, we must pay attention to the following. Spaceship cannot move unless the spaceship is independent of pressure gradient of vacuum field. No interaction is present between pressure gradient of vacuum field and spaceship. Spaceship does not move as long as the propulsion engine generates the pressure gradient or potential gradient in the surrounding area of spaceship, due to the interaction between pressure gradient of vacuum field and spaceship. This is because an action of propulsion engine on space is in equilibrium with a reaction from space. It is consequently necessary to shut off the equilibrium state to actually move the spaceship. As a continuum, the space has a finite strain rate, i.e., speed of light. When the propulsion engine stops generating the pressure gradient of vacuum field, it takes a finite interval of time for the generated pressure gradient of vacuum field to return to ordinary vacuum field. In the meantime, the spaceship is independent of pressure gradient of vacuum field. It is therefore possible for the spaceship to proceed ahead receiving the action from the vacuum field.

In general, a body cannot move carrying, or together with, a field that is generated by its body from the standpoint of kinematics. In other words, the body cannot move unless the body is independent of the field. This is because an action on the field and a reaction from the field are in the state of equilibrium.

As mentioned above, since the propulsion engine must necessarily be shut off for propulsion, the spaceship can get continuous thrust by repeating the alternate ON/OFF change in the engine operation at a high frequency.

Concerning the propulsion principle of field propulsion system, the distribution of field as shown in Figure 3 is fundamental; accordingly, several kinds of propulsion systems have been proposed. Even if any propulsion system is selected, whether the constituents of pressure gradient or potential gradient generated by propulsion engine are curvature, metric, zero-point radiation pressure or entropy, the propulsion principle of field propulsion system is the identical.

Further, as is already explained, all propulsion system based on the momentum thrust receives the reaction thrust by expelling the propellant mass. However, since no propellant is necessary for field propulsion, field propulsion is well called a propellant-less propulsion.

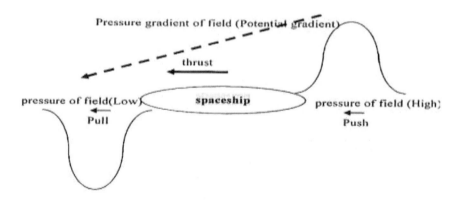

Figure 3. Fundamental propulsion principle of Field propulsion.

Basic Physical Concepts Applied for Field Propulsion

The propulsion principle of field propulsion assumes that space as a vacuum possesses a substantial physical structure. Field propulsion is propelled receiving the propulsive force (i.e., thrust) arises from the interaction of the substantial physical structure. The substantial physical structure is interpreted from both General Relativity in the view of macroscopic structure and Quantum Field Theory in the view of microscopic structure [5, 13].

Firstly, General Relativity is the geometric theory of gravitation and the gravitation is explained by curved space. The curvature of space plays an important role. Although, the curvature is quantity in mathematics, the curvature has relations with continuum mechanics such as expansion, contraction, elongation, torsion and bending. This physical relation indicates that the space as a vacuum can be considered as a kind of elastic body like rubber in continuum mechanics. Therefore, the propulsion system used General Relativity is to be proposed from the standpoint of continuum mechanics.

Secondly, space-time as a vacuum is generally viewed as a transparent and ubiquitous infinitive empty continuum, upon which physical events take place. However, Quantum Field Theory and quantum electrodynamics (QED) views it as possessing vigor and vitality over scales of time and space. Figure 4 shows the fine structure of space. Such vigor and vitality are the zero-point fluctuations of the vacuum electromagnetic field (vacuum perturbation) like Figure 4 (a), and the continuous creation and annihilation of virtual particle pairs. Further, according to the latest quantum optics, although until recently, it has been considered that the control of vacuum perturbation was utterly impossible, at present, it is proven that the vacuum perturbation can be controlled by squeezed light technology. At the present day, it is possible to increase the energy density locally above the vacuum state and vice versa, decrease the energy density locally below the vacuum state. That is, the squeezed light generates the squeezed vacuum states and yields the coordination geometry of energy density.

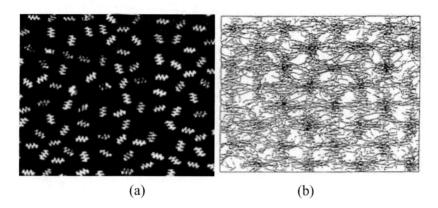

(a)   (b)

Figure 4. Fine structure of space.

Furthermore, the strings of superstring theory are considered as the threads of the space-time fabric.

String seems to be the fundamental element of the substructure or fine structure of space-time. Supposing that the string is the constituent of space-time is suggestive of the existence of possible quantum states for space-time. This indicates that the entropy of space-time can be defined as an assembling of strings. Strings as the constituents of space-time correspond to the polymer chains in the elastic body like Figure 4 (b). Since the statistical entropy is the logarithm of the number of states (i.e., degeneracy of system), it is necessary to consider what kinds of physical state exist.

Therefore, the propulsion system used Quantum Field Theory is to be proposed from the standpoint of quantum physics.

General Relativistic Field Propulsion System

On the supposition that space is an infinite continuum, continuum mechanics can be applied to the so-called "vacuum" of space. This means that space can be considered as a kind of transparent elastic field. That is, space as a vacuum performs the motion of deformation such as expansion, contraction, elongation, torsion and bending. The latest expanding universe theory (Friedmann, de Sitter, inflationary cosmological model) supports this assumption. We can regard space as an infinite elastic body like rubber. If space curves, then an inward normal stress "$-P$" is generated. This normal stress, i.e., surface force serves as a sort of pressure field (Figure 5).

$$-P = N \cdot (2R^{00})^{1/2} = N \cdot (1/R_1 + 1/R_2) , \qquad (5)$$

where N is the line stress, R_1, R_2 are the radius of principal curvature of curved surface, and R^{00} is the spatial curvature.

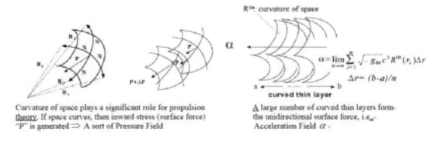

Figure 5. Curvature of Space.

It is now understood that the membrane force on the curved surface and each principal curvature generate the normal stress "$-P$" with its direction normal to the curved surface as a surface force. The normal stress "$-P$" is towards the inside of surface as shown in Figure 5. A thin-layer of curved surface will be taken into consideration within a spherical space having a radius of R and the principal radii of curvature which are equal to the radius ($R_1 = R_2 = R$). Since the membrane force N (serving as the line stress) can be assumed to have a constant value, Eq. (5) indicates that the curvature R^{00} generates the inward normal stress " $-P$" of the curved surface. The inwardly directed normal stress serves as a kind of pressure field. When the curved surfaces are included in a great number, some type of unidirectional pressure field is formed. A region of curved space is made of a large number of curved surfaces and they form the field of unidirectional surface force (i.e., normal stress). Since the field of surface force is the field of a kind of force, matter in the field is accelerated by the force, i.e., we can regard the field of surface force as the acceleration field. A large number of curved thin layers form the unidirectional acceleration field. Accordingly, the spatial curvature R^{00} produces the acceleration field α. Therefore, the curvature of space plays a significant role.

In detail, this concept is described in chapter 4 "Space Drive Propulsion: Typical Field Propulsion System".

Quantum Field Theoretical Propulsion System

In Quantum Field Theory, the fabric of space is visualized as consisting of fields, with the field at every point in space and time being a quantized simple harmonic oscillator, with neighboring oscillators interacting and one has a contribution of from every point in space. This suggests that the structure of space-time is also composed of some kinds of physical microstates and offers the properties of entropy.

The strings of superstring theory are considered as the threads of the space-time fabric. In a sense, it is as if individual strings are the "shards" of space-time, and only when they appropriately undergo sympathetic vibrations, the conventional notions of space-time emerge. Accordingly, string seems to be fundamental element of the substructures of space-time. This indicates that strings might behave like the polymer chains of some elastic bodies like rubber. In general, elasticity has two kinds of nature, that is, energy elasticity (crystalline elasticity) like spring and entropy elasticity (rubber elasticity) like rubber. Energy elasticity is due to the deformation of interatomic distance or displacement between molecules. It corresponds to the decrease of internal energy. Entropy elasticity (rubber elasticity) is due to the thermal motion of the polymer chains. It corresponds to an increase of entropy. Elasticity of rubber is very different from that of crystalloid solids. The elastic constant of rubber increases with temperature.

The space as vacuum is considered to preserve the properties of entropy elasticity from the viewpoint of the latest cosmology and theory of elementary particles. Further, the space as vacuum is considered that the entropy of space-time can be defined as an assembling of strings, and strings as the constituents of space-time correspond to the polymer chains in the elastic body. As shown in the usual rubber elasticity, the elastic force is induced by entropy gradient in the direction of increasing entropy (from small entropy to large entropy). So the elastic force "F" from the field of entropy gradient is generated.

The excited space shows the property of rubber elasticity, and as is well known in the statistical theory of rubber elasticity, the elastic force F is given by

$$F = T\frac{\partial S}{\partial r}, \quad S = k\log W, \quad \Rightarrow \quad F = kT\frac{\partial \log W}{\partial r}, \qquad (6)$$

where T is the temperature density as energy density, S is the entropy, r is the distance, W is the number of microstates, k is the Boltzmann's constant.

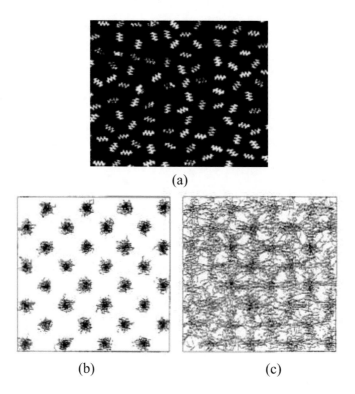

Figure 6. Zero-Point Fluctuation of Electromagnetic Energy in a Vacuum and Fine Structure of Space.

Figure 6 shows Zero-Point Fluctuation of Electromagnetic Energy in a Vacuum and Fine Structure of Space. Supposing the entropy structure of space-time, since the statistical entropy is the logarithm of the number of states (i.e., degeneracy of system), it is necessary to consider what kinds of physical state exist. Figure 6 (a) shows Zero-Point Fluctuation. Figure 6 (b) and Figure 6 (c) show that the open strings cling to the field of space. Figure 6 (b) shows the state of present cosmic space in ultra-low temperature, and Figure 6 (c) shows the state of the early universe in ultra-high temperature. The excitation of space implies that the ordered phase of open strings clung to space in Figure 6 (b) is transferred to the disordered phase of open strings clung to space in Figure 6 (c) by some trigger. It corresponds to that the number of twined open strings is transferred from ordered phase (small entropy) to disordered phase (large entropy). This picture indicates that these states can be interpreted as entropy.

Figure 7 shows the propulsion method by the density difference of the zero point oscillator. How to place the number of zero point oscillators in front of and behind the spaceship in a non-equilibrium manner is the function of the engine. For the front of spaceship, suppose that the number of zero point oscillator is small and the zero point radiation pressure is small in a normal vacuum condition. If the number of zero point oscillator is increased to some extent behind the spaceship, the zero point radiation pressure increases.

The spaceship will move forward by being pushed from the back vacuum with large zero point radiation pressure to the forward vacuum with low zero point radiation pressure. Just like a ping-pong ball underwater is pushed in a direction of low water pressure from a higher water pressure.

At the present time, although the empirical Casimir forces cannot be directly applied to propulsion, Casimir effects is the sole theory and experiment which indicates the possibility of extracting thrust from vacuum. In order to extract the micro-thrust from vacuum as a first stage, the perturbation of vacuum and the generation of localized inhomogeneous field are indispensable. The vacuum fluctuates with oscillations of the electric field, and these vacuum fluctuations and the zero-point energy have a

Introduction to Field Propulsion

common origin in the quantized vacuum. The vacuum fluctuations are considered as quantum noise. Until recently, it has been considered that the control of vacuum perturbation was utterly impossible. However, at present, it is proven that the vacuum perturbation can be controlled by squeezed light technology. Therefore, it is possible to increase the energy density locally above the vacuum state and, vice versa, decrease the energy density locally below the vacuum state. Since the energy density is equivalent to the pressure, the control of energy density is to control zero-point radiation pressure. So, the squeezed light generates the squeezed vacuum states and yields the coordination geometry of energy density, i.e., zero-point radiation pressure. One way to realize the zero point oscillator arrangement as shown in Figure 7 is to use a squeezed laser.

The theoretical possibility of extracting thrust from the excited vacuum (i.e., squeezed vacuum state) induced by the control of squeezed light and the experimental concepts are described next in detail.

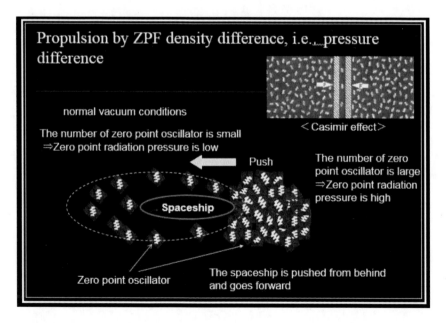

Figure 7. Propulsion by ZPF density.

Squeezed Light

According to the Quantum Optics [14, 15, 16], it is well known that the quantized field fluctuates; the zero-point energy and the vacuum fluctuations actually present severe problems in Quantum Field Theory. The zero-point vacuum fluctuations actually give rise to observable effects, such as spontaneous emission of light, the Lamb shift and the Casimir effect. Although the light generated by laser is in a coherent state and its fluctuation (i.e., quantum noise) is exactly the same as for the vacuum, the squeezed light has less quantum noise in one of the quadratures than for a coherent state or a vacuum state; the fluctuations in that quadrature are squeezed. In this sense, squeezed light contains phase-dependent noise, reduced below that of the vacuum for some phases and enhanced above that of the vacuum. These squeezed states of light have been extensively investigated recently in quantum optics and have been experimentally realized also in Japan. To mathematically generate squeezed light, it is through the action of a "squeeze" operator defined as:

$$\hat{S}(\zeta) = \exp\left[\frac{1}{2}\zeta^* \hat{a}^2 - \frac{1}{2}\zeta(\hat{a}^+)^2\right], \quad \zeta = re^{i\theta}, \tag{7}$$

where r is known as the squeeze parameter and $0 \leq r < \infty$ and $0 \leq \theta \leq 2\pi$. When $|\zeta| = r = 0$, the squeezed state reduces to coherent states. Here, $\hat{a}$ and $\hat{a}^+$ are the annihilation and creation operators. The squeeze operator $\hat{S}(\zeta)$ acting on the vacuum state (denoted as |0>) would create the squeezed vacuum state (denoted as |0,ζ>):

$$|0,\zeta\rangle = \hat{S}(\zeta)|0\rangle. \tag{8}$$

A more general squeezed state for a single mode can be obtained by applying the displacement operator $\hat{D}(\alpha)$ as:

$$|\alpha,\zeta\rangle = \hat{D}(\alpha)\hat{S}(\zeta)|0\rangle, \quad \alpha = |\alpha|e^{i\theta} \tag{9}$$

Introduction to Field Propulsion

that is, the squeezed state $|\alpha, \zeta\rangle$ is obtained by first squeezing the vacuum and the displacing it. By displacing the vacuum state $|0\rangle$, the coherent states are given as: $|\alpha\rangle = \hat{D}(\alpha)|0\rangle$.

When $\alpha = 0$, it is known as a squeezed vacuum state. Such a state is not the vacuum state so long as $\zeta \neq 0$, but rather a superposition of states containing even numbers of particles. The variances of two quadratures of the squeezed vacuum state for $\theta = 0$ are given by following Eq. (10).

For $\theta = 0$, evidently squeezing exists in the $\hat{X}_1$ quadrature. For $\theta = \pi$, the squeezing will appear in the $\hat{X}_2$ quadrature.

$$\left\langle (\Delta \hat{X}_1)^2 \right\rangle_{sq-vac} = \frac{1}{4} e^{-2r}, \quad \left\langle (\Delta \hat{X}_2)^2 \right\rangle_{sq-vac} = \frac{1}{4} e^{2r}. \quad (10)$$

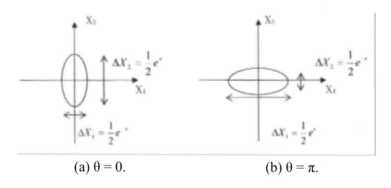

(a) $\theta = 0$. (b) $\theta = \pi$.

Figure 8. Error Ellipse for a Squeezed Vacuum State.

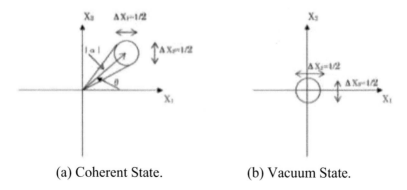

(a) Coherent State. (b) Vacuum State.

Figure 9. Phase-Space Portrait (Non-Squeezed State).

A representation of the squeezed vacuum state for θ = 0, where the fluctuations in X_1 are reduced, is given in Figure 8 (a), whereas for θ = π, where the fluctuations in X_2 are reduced, is given in Figure 8 (b). In the cases with θ = 0 or θ = π, the squeezing is along either $\hat{X}_1$ or $\hat{X}_2$. A "squeezed state" is a state whose fluctuations in one quadrature phase are less than zero-point fluctuations (or the fluctuations in any coherent state), and whose fluctuations in the other phase are larger than zero-point fluctuations. Namely, the squeeze operator attenuates one component of the amplitude, and it amplifies the other component. The degree of attenuation and amplification is determined by squeeze parameter r = |ζ|. A coherent state and a vacuum state where in both cases the fluctuations of the quadrature operators are equal, $\Delta X_1 = \Delta X_2 = 1/2$ (see Figure 9 (a), (b).). As is shown above, the squeezed state |α,ζ> has the same expected complex amplitude as the corresponding coherent state |α>, and it is a minimum-uncertainty state for X_1 and X_2.

The difference lies in its unequal uncertainties for X_1 and X_2. That is, squeezed states are less noisy (or larger noisy) than for a field in a vacuum state. This implies that ZPF noise (quantum noise) can be manipulated.

For a reference, the graph of squeezed vacuum state of electric field versus time is shown as Figure 10. Although the fluctuation in coherent state is temporally constant, the fluctuation in squeezed state periodically varies from the maximum value of εe^{r} to the minimum value of εe^{-r} [17]. Here, $\varepsilon = (\hbar\omega/2\varepsilon_0 V)^{1/2}$, ε_0 is dielectric constant of vacuum, $\hbar$ is Planck Constant (1.054 × 10^{-34} J.s), ω is angle frequency (rad/s) and V is volume (m³).

The graphs of electric field versus time for three states of the electromagnetic field are shown as Figure 11 $^{(\alpha \neq 0)}$. For a coherent state, the rotation of the error circle leads to a constant value for the variance of the electric field (Figure 11(a)). For a squeezed state, the rotation of the error ellipse leads to a variance that oscillates with frequency 2ω (Figure 11(b), (c)).

Introduction to Field Propulsion 25

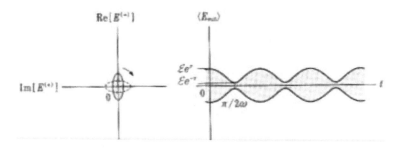

Figure 10. Electric Field for Squeezed Vacuum State (Matsuoka, 2000).

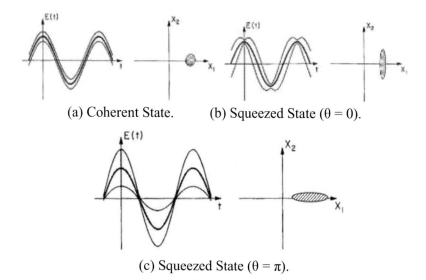

(a) Coherent State. (b) Squeezed State ($\theta = 0$).

(c) Squeezed State ($\theta = \pi$).

Figure 11. Electric Field for Coherent & Squeezed State.

Figure 12 shows the squeezed vacuum state. Vacuum state line (($\overline{}$): ΔE ; pink) indicates non-squeezed usual vacuum state (coherent laser beam), and the fluctuations in the electric field ΔE is constant at all times. The fluctuations in the electric field ΔE are the same value both interior of laser beam and exterior of laser beam. The fluctuations in the electric field ΔE exterior of laser beam are usual vacuum state. On the contrary, concerning the squeezed vacuum state induced by squeezed light, the fluctuation in squeezed state periodically varies from the maximum value of εe^{r} to the minimum value of εe^{-r} ; $\varepsilon = (\hbar \omega / 2\varepsilon_{0} V)^{1/2}$ (see Figure 12 (a)).

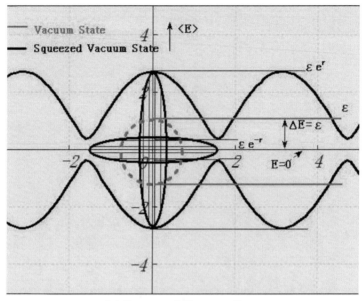

(a) Squeezed Vacuum State

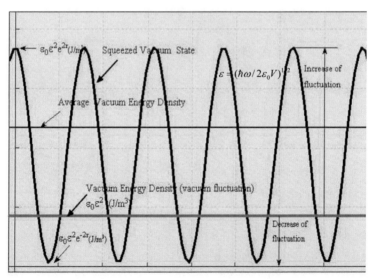

(b) Squeezed Vacuum Energy Density

Figure 12. Electric Field for Squeezed Vacuum State.

Next, let us consider about vacuum energy density shown in Figure 12 (b).

As mentioned previously, non-squeezed vacuum energy density ($\varepsilon_0 \varepsilon^2$: pink line) is constant value of $\varepsilon_0 \varepsilon^2 = \varepsilon_0 (\sqrt{\hbar \omega / 2\varepsilon_0 V})^2 = \hbar \omega / 2V \, (J/m^3)$. In the case of large value of squeeze parameter "r", the energy density of vacuum increases exponentially in accordance with $\varepsilon_0 \varepsilon^2 e^{2r} = \hbar \omega e^{2r} / 2V \, (J/m^3)$. The fluctuations in squeezed vacuum state exhibits larger fluctuations, and hence more energy density, than a non-squeezed vacuum state. On the other hand, the fluctuations in squeezed vacuum state also exhibits smaller fluctuations, and hence less energy density, than a non-squeezed vacuum state. Since the fluctuation in squeezed state periodically varies from the maximum value of εe^r to the minimum value of εe^{-r}, there occurs the oscillatory energy by periodic occurrences of both larger and smaller fluctuations compared to the non-squeezed vacuum state. Consequently, the average vacuum energy density in squeezed state (blue line) increases as compared to non-squeezed vacuum state (pink line). Thus, the squeezed light can alter the energy density of vacuum, that is, vacuum perturbation (ZPF) can be controlled by squeezed light technology. As a result, the vacuum state develops structure in the sense that the energy density is no longer spatially homogeneous.

Vacuum Perturbation Induced by Squeezed Light

The energy density for the squeezed vacuum state was obtained referring to Weigert's article [18]. The following description is based on the cavity which has fixed boundaries, and is focused on the states of the field which minimize the expectation value of the energy in a prescribed region. The spatial variation of the energy density $\hat{W}(x)$ associated with kth mode is distributed inhomogeneously in the cavity as the following:

$$\langle \sigma_k | \hat{W}(x) | \sigma_k \rangle = \frac{\hbar \omega_k}{2LS} \left(\mu_k \sin^2 \omega_k (x + \Lambda) + \frac{1}{\mu_k} \cos^2 \omega_k (x + \Lambda) \right). \quad (11)$$

Each mode is a squeezed vacuum state which is denoted by $|\sigma_k\rangle$ since its degree of squeezing is determined by the asymmetry parameter σ_k. Assuming that the Casimir-type cavity, the energy density U near the plate in the state $|\sigma_k\rangle$ (that is, $x = \lambda$) is obtained from Eq. (11), using $\omega_k = k\pi c/L$, considering a cavity of length $L = \Lambda + \lambda$.

Its boundaries at $x_- = -\Lambda$, $x_+ = \lambda$ are assumed [18]. L is the length of cavity; S is the area of squeezed light.

$$U = \langle \sigma_k | \hat{W}(\lambda) | \sigma_k \rangle = \frac{1}{\mu_k} \frac{\hbar \omega_k}{2LS} = \frac{\hbar \omega_k}{2LS} e^{2r} = \frac{1}{\mu_k} \frac{\hbar k \pi c}{2L^2 S} = \frac{\hbar k \pi c}{2L^2 S} e^{2r} \quad (J/m^3).$$

(12)

By the way, the relation between the squeeze parameter r and μ_k is given by:

$$\mu_k = e^{-2r}.$$ (13)

On the other hand, the squeeze parameter r is given by (cgs units) [19]:

$$r = \chi \left(\frac{4\pi \omega_s \ell}{c n_s} \right) |E_p| = \chi \left(\frac{4\pi \omega_s \ell}{c n_s} \right) \left(\frac{8\pi P_p}{c n_p A} \right)^{1/2}.$$ (14)

Here, χ is the effective nonlinear susceptibility, E_p is the amplitude of the pump wave's electric field, ℓ (cm) is the length of the nonlinear medium, n_s and n_p are the index of refraction at the signal and pump frequencies, P_p (erg/s) is the pump power distributed over an area $A(cm^2)$.

From Eq. (12), energy density U(z) is the function of squeeze parameter "r". Here, we consider that we control the squeeze parameter in accordance with the position of z axis (see Figure 13). Therefore, the squeeze parameter "r" is the function of "z". For simplicity, setting $r(z) = \alpha z$, further, supposing that $\alpha = 1(1/m)$, then, $r(z) = z = r$. This "α" is not "α" denoted in Eq. (9).

Accordingly, we get the force F induced by squeezed light from Eq. (12), which generates a vacuum perturbation in the squeezed state as follows:

$$F = -\frac{\partial U(z)}{\partial z} = -\frac{\partial}{\partial z}\left(\frac{\hbar k \pi c}{2L^2 S}e^{2\alpha}\right) = -\frac{\alpha \hbar k \pi c}{L^2 S}e^{2r} = -\frac{\alpha \hbar k \pi c}{L^2 S}\exp$$
$$\left(2\chi\left(\frac{4\pi\omega_s \ell}{cn_s}\right)\left(\frac{8\pi P_p}{cn_p A}\right)^{1/2}\right) (N/m^3). \quad (15)$$

Now, let us consider the squeezed vacuum energy density, from another standpoint. Kuo, Ford and Davis showed that the squeezed quantum states generate the negative energy density, in their article [20, 21]. In a squeezed vacuum, $r \neq 0$, $\alpha = 0$ in Eq. (9), also taking $\theta = 0$, the expectation value of the stress tensor is given by:

$$\langle 0,\zeta|T_{00}|0,\zeta\rangle = \frac{\hbar\omega}{L^3}\sinh r[\cosh r \cos 2\theta + \sinh r]. \quad (16)$$

Here, T$_{00}$ is the energy density component of stress tensor T$_{\mu\nu}$ in gravitational field.

In the case of large squeeze parameter "r" as shown in Figure 14, the energy density accords with described above equation Eq. (12), taking notice of S = L^2. In addition, considering photon number in squeezed vacuum, the expectation value of photon number is obtained as the following:

$$\langle \hat{n} \rangle = \sinh^2 r. \quad (17)$$

The energy of one photon is $\hbar\omega$, so the energy can be obtained by Eq. (12), and then, energy density is obtained like Figure 14.

$$U_{vac} = \hbar\omega\langle \hat{n}\rangle = \hbar\omega\sinh^2 r \approx \frac{\hbar\omega}{4}e^{2r} (J). \quad (18)$$

This accords with described above equation Eq. (12) approximately. In any case, a common result is that the vacuum energy density in squeezed state locally increases exponentially as value of squeeze parameter increases in accordance with e^{2r}.

Next, let us consider propulsion principle again. The propulsion principle described here akin to nuclear propulsion system as shown in Figure 13. As is well known, nuclear propulsion is propelled by receiving the action of blast wave of pellet. Similarly, it may be possible to consider that starship is propelled by receiving the action of blast of vacuum perturbation induced by extremely increased vacuum energy density, i.e., strong zero-point radiation pressure. This strong zero-point radiation pressure is periodically generated as impulse drive by pulsed squeezed light [22, 23].

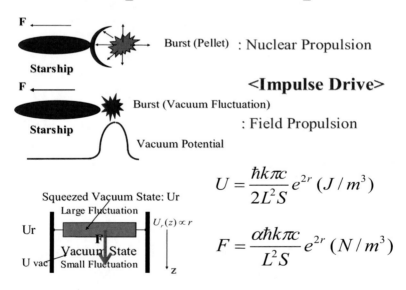

Figure 13. Propulsion Principle & Impulse Drive.

Squeezed Vacuum Energy Density

- **Kuo and Ford;**

$$\langle 0,\zeta|T|0,\zeta\rangle = \frac{\hbar\omega}{L^3}\sinh r[\sinh r + \cosh r \cos 2\theta]$$

$$\sinh r = \frac{e^r - e^{-r}}{2} \cong \frac{e^r}{2}, \quad \cosh r = \frac{e^r + e^{-r}}{2} \cong \frac{e^r}{2}, \quad \because e^r \gg e^{-r}.$$

$\cos 2\theta = 1$ and $r \gg 1$, then, $e^r \gg e^{-r}$

$$\langle 0,\zeta|T|0,\zeta\rangle = \frac{\hbar\omega}{L^3}\sinh r[\sinh r + \cosh r \cos 2\theta] = \frac{\hbar\omega}{L^3}\frac{e^r}{2}\left[\frac{e^r}{2}+\frac{e^r}{2}\right] = \frac{\hbar\omega}{L^3}\frac{e^{2r}}{2} = \frac{\hbar\omega}{2L^3}e^{2r}$$

- **Photon Number in Squeezed Vacuum;** $\langle \hat{n}\rangle_{r\neq 0} = \sinh^2 r$

$$Energy_{vac} = \hbar\omega\langle\hat{n}\rangle = \hbar\omega\sinh^2 r = \hbar\omega\left(\frac{e^r - e^{-r}}{2}\right)^2$$

$$\approx \frac{\hbar\omega}{4}e^{2r}(J), \text{ then, } U_{vac} = \frac{\hbar\omega}{4L^3}e^{2r}(J/m^3).$$

Figure 14. Squeezed Vacuum Energy.

High-Power Squeezed Laser and Experimental Plan

From here, we would like to show the basic experiment plan. Squeezed state is produced by parametric amplifier. The experimental small chip is connected to hanging wire such as pendulum. When the squeezed vacuum state is instantaneously generated by pulsed squeezed light, the experimental chip may swing receiving the action of blast of vacuum perturbation induced by extremely increased vacuum energy density, i.e., strong zero-point radiation. Figure 15 shows the basic principle of generating squeezed state by parametric amplifier. A degenerate parametric amplifier consists of a second-order nonlinear crystal pumped by an intense laser beam at angular frequency $\omega_P = 2\omega$. A weak signal beam at angular frequency $\omega_S = \omega$ is also introduced. The nonlinear crystal mixes the signal with the pump and produces an idler beam at angular frequency $\omega_I = \omega$. The pumped nonlinear crystal acts as a phase-sensitive amplifier for signal modes at angular frequency ω.

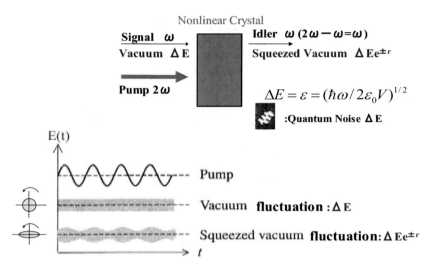

Figure 15. Parametric Amplifier.

We assume that there is no signal beam present at the input of the crystal. In this case, the signal is taken from the ever-present vacuum modes. The vacuum modes consist of a randomly fluctuating field of average amplitude. The nonlinear process either amplifies or de-amplifies the vacuum depending on its phase. This produces an output field as shown in squeezed vacuum. That is, with no signal input, the nonlinear crystal amplifies and de-amplifies the vacuum modes, hence producing squeezed vacuum states. As previously mentioned, although the fluctuation in a coherent state or vacuum state is temporally constant, the fluctuation in a squeezed vacuum state periodically varies from the maximum value of εe^{r} to the minimum value of εe^{-r}. The squeezed light can alter the energy density of the vacuum. The squeezed vacuum exhibits smaller fluctuations, and hence less energy density, than the vacuum in space-time regions. On the other hand, the squeezed vacuum also exhibits larger fluctuations, and hence more energy density, than the vacuum in space-time regions. That is, the oscillatory energy density by periodic occurrences of both smaller and larger fluctuations compared to the unsqueezed vacuum is generated in the cavity. Therefore, the energy density of the vacuum increases exponentially as value of squeeze parameter (r) increases in accordance with "εe^{r}". Inversely, the energy density of the vacuum decreases exponentially as value of squeeze parameter (r) increases in accordance with "εe^{-r}" and is confined within the limits of vacuum value.

Figure 16 shows the block diagram to generate the squeezed light for experiment. An Nd:YAG laser system with a second harmonic generator (SHG) produces a powerful 20-W laser beam (thick solid line; 2ω). A powerful 20-W laser beam is bypassed by dichroic mirror. The laser beam incident on an Optical Parametric Amplifier (OPA) through beam splitter, PBS (Polarized Beam Splitter for 2nd harmonics) and f# = 10 lens to achieve 10-MW/cm² laser intensity inside the OPA for efficient conversion to signal and idler beams, denoted by thin solid and dashed lines. Both beams are reflected by a curved mirror whose curvature and position are designed to output a collimated beam through the f#10 lens. A low power fundamental frequency beam from the laser system is redirected by the PBS, reflected by a mirror and incident to squeezed light detector where it serves as a probe beam for homodyne detection of squeezing. The power of the probe is adjusted by a λ/2 plate. The phase of the probe beam is modulated by a piezo actuator to see phase dependence of squeezing. Signal and idler beams are also incident on the detector.

The possibility of vacuum perturbation induced by squeezed light and its theoretical equation are obtained. The value of micro-thrust extracted from quantum vacuum depends on the squeeze parameter. The large value of squeeze parameter can be achieved by high-power laser. However, it became clear that the squeezed light generated by high intensity laser (MW-GW) was capable of achieving vacuum perturbation large enough to observe the extraction of micro-thrust (e.g., a few m N). The energy of one pulse shall be high energy (e.g., 1mmJ to 10J per pulse). So, the ultrahigh-intensity laser (TW-PW) such as CPA (Chirped Pulse Amplification) laser is not necessarily.

Above all, the most important technical future subject is the development of nonlinear crystal having the large value of second-order nonlinear susceptibility for the enhancement of squeeze parameter "r" and so on.

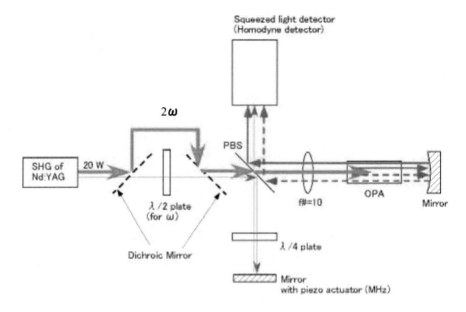

Figure 16. Laser System for High Power Squeezed Light.

The proposed way to obtain thrust appears to be based on *pulsed* squeezed light. Thus, the space vehicle should have a high-power source onboard.

Other Method for Propulsion

Figure 7 shows the propulsion method by the density difference of the zero point oscillator. As shown in Figure 7, a squeezed laser was used as a method of densely arranging the zero point oscillator on the back of the spaceship. However, other method may exist; for example, inside a large tapered metal cylinder closed at both ends, by reflecting back and forth inside the cylinder using oscillated microwave, it may be possible that the zero point oscillator is moved in one direction and concentrated locally behind the spaceship.

REFERENCES

[1] Minami Y., "Space Strain Propulsion System", *16th International Symposium on Space Technology and Science (16th ISTS)*, Vol.1, 1988: 125-136.

[2] Minami, Y., "An Introduction to Concepts of Field Propulsion", *JBIS* 56, 2003: 350-359.

[3] Minami, Y., "Possibility of Space Drive Propulsion", In *45th Congress of the International Astronautical Federation (IAF)*, (IAA-94-IAA.4.1.658), 1994.

[4] Minami, Y., "Spacefaring to the Farthest Shores-Theory and Technology of a Space Drive Propulsion System", *Journal of the British Interplanetary Society (JBIS)* 50, 1997: 263-76.

[5] Minami, Y., "A Superstring-Based Field Propulsion Concept", *JBIS*, 57, 2004: 216-224.

[6] Puthoff, H.E., "Gravity as a zero-point-fluctuation force", *Phys. Rev. A*, **39**, pp.2333-2342, 1989.

[7] Haisch, B. Rueda, A. and Puthoff, H.E. "Inertia as a zero-point-field Lorentz force", *Phys. Rev. A*, **49**, pp.678-694, 1994.

[8] Millis M.G., "Exploring the Notion of Space Coupling Propulsion, In Vision 21: Space Travel for the Next Millennium", *Symposium Proceedings, Apr 1990, NASA-CP-10059*, 1990: 307-316.

[9] Froning Jr. H.D., "Vacuum Energy For Power and Propulsive Flight?", *AIAA 94-3348, 30th AIAA/ASME/SAE/ASEE Joint Propulsion Conference*, June 27-29, 1994/Indianapolis, IN, USA.

[10] Froning Jr. H.D., Barrett T.W., "Inertia Reduction-And Possibly Implusion-By Conditioning Electromagnetic Fields", *AIAA 97-3170,33rd AIAA/ASME/SAE/ASEE, Joint Propulsion Conference & Exhibit*, July 6-9, 1997.

[11] Puthoff II.E., Little S.R., Ibison M., "Engineering the Zero-Point Field and Polarizablel Vacuum for Interstellar Flight", *Journal of The British Interplanetary Society*, Vol.55, 2002: 137-144.

[12] Millis M.G., Williamson G. S., (Eds.), "NASA Breakthrough Propulsion Physics Workshop Proceedings", Jan 1999, *NASA-CP-1999-208694*: 263-273.

[13] Minami, Y., Musha, T., "Field propulsion systems for space travel", *Acta Astronautica*, 82, 215-220 (2013).

[14] Gerry, C.C. and Knight, P.L., *Introductory Quantum Optics*, Cambridge University Press, 2005.

[15] Vedral, V., *Modern Foundations of Quantum Optics*, Imperial Collage Press, London, 2005.

[16] Walls, D.F. and Milburn, G.J., *Quantum Optics*, Springer-Verlag, Berlin Heidelberg, 1994.

[17] Matsuoka, M., *Quantum Optics*, Shokabo, Tokyo Japan, 2000.

[18] Weigert, S., "Spatial squeezing of the vacuum and the Casimir effect", *Phys. Lett A.*, 214, pp.215-220, 1996.

[19] Caves, C.M., "Quantum-mechanical noise in an interferometer", *Phys. Rev. D*, 23, pp.1693-1708, 1981.

[20] Kuo, C-I. and Ford, L.H., "Semiclassical gravity theory and quantum fluctuations", *Phys. Rev. D*, 47, pp.4510-4519, 1993.

[21] Davis, E.W. and Puthoff, H.E., "Experimental Concepts for Generating Negative Energy in the Laboratory", in proceedings of Space Technology and Applications International Forum (STAIF-2006), edited by M. S. El-Genk, *AIP Conference Proceedings*, Melville, New York, 2006, pp.1362-1373.

[22] Minami Y., "Extraction of Thrust from Quantum Vacuum Using Squeezed Light", STAIF-2007, edited by Mohamed S. El-Genk, *AIP Conference Proceedings*, Feb.11-15, 2007, Albuquerque, NM, USA.

[23] Minami, Y., "Preliminary Theoretical Considerations For Getting Thrust Via Squeezed Vacuum", *JBIS*, 61, 2008: 315-321.

Chapter 2

ENERGY FOR SPACEFLIGHT POWER AND PROPULSION FROM SPACE ITSELF

Herman D. Froning Jr.

2.1. INTRODUCTION

I was only mildly interested in spaceflight when it began. Space was awesome, but also hostile, cold, air-less and filled with lethal solar-cosmic radiation, with no inhabitable world like our own near enough to reach in a human life (Figure 1). Then I stumbled upon a translation of a paper by a German rocket scientist Eugen Sanger [1] showing that a ship moving at almost light speed c would, in effect, contract space ahead of itself and slow time inside itself in accord with Special Relativity (SR). And this space contracting and time slowing would increase more and more as ship speed became nearer and nearer to c. Using one Earth gravity acceleration and deceleration during first and final parts of trips, Sanger, showed the center of our Milky Way galaxy would take 30,000 years of Earth time to reach, but only 19 years would elapse on the ship. A longer trip to the Andromeda M-31 galaxy would take 2.2 million Earth years, but only 28 years on the ship.

Figure 1. Interstellar Gas Clouds.

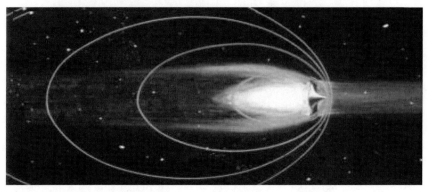

Figure 2. Interstellar ramjet flight.

And he showed that the universe's enormous circumference (then believed 3.0 billion light-years long) could be traveled in about 45 years of ship time - well within on-board human lives. So Sanger gave me hope that humanity might not be forever confined within our very small current part of the Cosmos.

Sanger assumed an ideal "photon rocket" which converted matter into energy by mutual annihilation of electrons and positrons and he assumed no structural mass. But even with a perfect rocket with no dry mass, his ships needed prohibitive amounts of fuel for even the shortest trips to the nearest stars. So, I soon concluded that too much fuel had to be carried by relativistic rocket ships. After some pondering I concluded that a possible source of starship fuel might be the denser clouds of neutral and ionized interstellar hydrogen inside nebulas (like the Horse-head nebula shown below) at densities as high as 10,000 atoms/cm^3.

Being familiar with missiles with rocket lower stages and ramjet upper stages I envisioned a Sanger-like anti-matter lower stage rocket for initial boost to almost speed-of-light c. Then most of the long interstellar ramjet flight would be flown as shown below by a ramjet-like stage that ingested and compressed the very long swath of hydrogen atoms in the gas clouds along the ship's long interstellar path through space (Figure 2). Then matter and fusion energy release from ingested atoms would be exhausted from the vehicle. And the resulting increase in the ship's momentum would accelerate the ship nearer and nearer to the speed of light c.

But I soon found that a distinguished fusion physicist, Dr. Robert Bussard [2] had thought of an interstellar ramjet concept years before than I had. Bussard's concept required magnetic funnel-ling of much more volumes of interstellar hydrogen than mine did. And many experts refuted the feasibility of magnetically funneling vast volumes of tenuous gases into his craft. Finally, though my ramjet needed less funneling, I abandoned interstellar gases.

2.2. QUANTUM FIELD ENERGY OF VACUUM

Shortly thereafter a company physicist referred me to the work of Professor John Wheeler of Princeton University. Wheeler showed the astonishing possibility that what seems to be inert and empty space is actually teaming with vigor and vitality over microscopic scales of time and distance, "fluctuating at the scale of the "Planck length" between configurations of varied curvature and topology" [3]. And these quantum fluctuations resulted in energy fluctuations occurring over very short times and distances, as shown below, throughout all cosmic space (Figure 3).

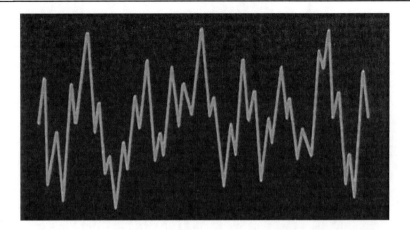

Figure 3. Fluctuation in zero-point energy state of EM quantum vacuum in small region of empty space.

Quantum energy fluctuations occurring in a given space can never be exactly known. But [3] showed "expectation values" of fluctuation energy and energy density in a cubical region of length L would be hc/L joules and hc/L^4 joules/cm^3 (where h is Planck's constant). And a nanometer region (about 10 times larger than an atom's diameter) was deemed the shortest distance over which an interaction with vacuum energy structure in a given region of quantum vacuum could be achieved. And within a nanometer length region, energy of 10^{-7} joules and energy density is 10^2joules/cm^3 could be expected. Shown below is a short part of the very long swath of quantum zero-point fluctuation energies that the frontal area of a ramjet-like ship would sweep-out and ingest along its very long interstellar route. And here, quantum zero-point energy fluctuations, like electromagnetic energy pulsations, would be somewhat like a very great many different lightning flashes of many different wavelengths or frequencies. So, lower-energy quantum fluctuations that are to be swept out are symbolized by cooler colors (such as blue and yellow and green) while the higher-energy fluctuations that are to be swept out are symbolized by warmer colors (such as orange and pink and white) as shown in Figure 4.

Energy for Spaceflight Power and Propulsion ... 41

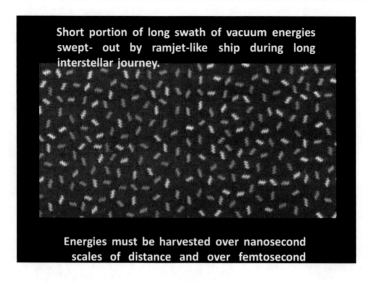

Figure 4. Quantum zero-point energy fluctuations.

Actual existence of the quantum vacuum has been confirmed by two discoveries. One is the "Lamb Shift" in the ground state of the hydrogen atom (when an alpha particle is emitted from the atom's nucleus during its occasional interaction with quantum vacuum. Another is the Casimir-Polder force predicted by Hendrick Casimir and Dirk Polder. This predicted force is caused by exclusion of vacuum fluctuations from the narrow cavity formed between 2 very closely-spaced conducting plates. This exclusion of vacuum fluctuations was predicted to result in more inward-pushing than outward-pushing vacuum pressure on the plates. This has subsequently been confirmed by numerous experiments performed over many years.

As shown in the representation on the next page, I used Bussard's ramjet representation of an interstellar ramjet's relativistic interaction with matter to make a similar one for a "quantum interstellar ramjet". This allowed gathered energy and available ship thrust to be calculated parametrically as a function of the efficiency that zero-point energy (zpe) could be extracted from the vacuum and converted into useful kinetic energy within the ship's exhaust. And such calculations were made for different scales of distance (L) over which given expectation values of zpe could be extracted from the vacuum. Beginning with such a representation, I was able to conclude that vacuum zpe extraction had to occur over nanometer scales of distance and within femto-second scales of time (Figur3 5).

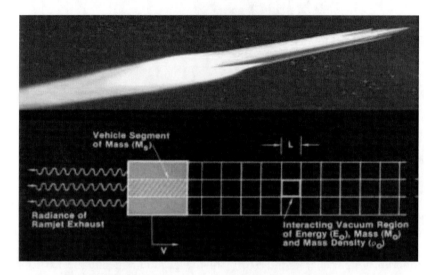

Figure 5. Quantum interstellar ramjet's relativistic interaction.

Figure 6. 2.2 Million Light-Year Journey to M31 Galaxy in less than 30 years of time on-board a ship.

A Quantum Interstellar Ramjet of Saturn 5 moon rocket size and mass was configured for very long inter-galactic flights to distant galaxies like the Andromeda nebula shown below in 29 ship years (Figure 6). And the ship's 10 meter diameter intakes ingested thousands of terra-coulombs of quantum fluctuation energy during the long 2.2 million light-year distance, as the ship developed many terra-watts of power in accelerating extremely close to the speed-of-light.

I first described the possibility of interacting with quantum fluctuation energies and of some requirements for harvesting the energies for power and thrust at a meeting of the British Interplanetary Society [4] in London in 1980. My paper "Requirements for a Quantum Interstellar Ramjet" probably did not raise very high interstellar flight hopes because it mainly revealed the extremely difficult problem of interacting with quantum fluctuations over nano-meter distances and within times of the order of femto-seconds. For, in 1980, such short times and distances were thousands of times less than those the most advanced laser and microwave devices could then probe. But they are now being approached by very advanced academic and government advanced petawatt pulsed lasers with sub-micrometer spot-sizes.

Quantum Field Energy for the Renewing of Matter

As you might imagine, after all my past explorations of possible ways to harvest the quantum vacuum's energies for spaceflight. I continue to hope someone someday will discover how to materialize concentrated quantum energies out of vacuum regions similar in size to those occupied by atoms. But Professor Derek Leinweber [5] at the University of Adelaide reveals that concentrated quantum energies emerge-from and return-to regions of space that are millions of times smaller than atoms. These quantum energies manifest themselves within any proton or neutron-occupied region of cosmic space within any region in the cosmos that contains 3 quarks. This particular vacuum is called the "Quantum Chromo Dynamic (QCD) Vacuum". It is associated with the Strong Force interaction that occurs within atomic nuclei containing protons and those containing protons and neutrons. Leinweber's computational grid below shows the beginning of a dynamic QCD vacuum interaction that has probably been going-on since matter was first formed in the "big-bang" beginning of our universe (Figure 7).

This interaction is extremely important as its energetics (that occurs trillions of times a second in every proton-occupied region of cosmic space) create about 97% of the mass-energy of our universe's seeable matter. Figure 8 shows a snapshot of a typical QCD vacuum interaction going on in Leinweber's QCD computational domain. This domain, which also contains computed patterns, masses and fields of very small "quarks" are also computed, but not shown in this particular figure. Here, locations, and masses of tiny, un-seen quarks continually-suddenly-swiftly change as mass-less energy structures, called "gluons" or "gluon fields" materialize out of QCD quantum vacuum - expanding and twisting and shrinking and then dematerializing back into the QCD vacuum in less than about a trillionth of a second.

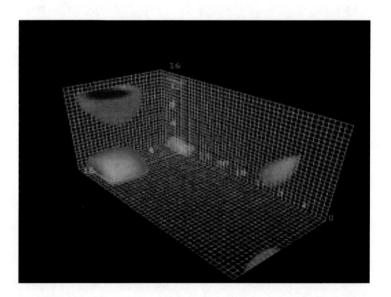

Figure 7. Quantum Chromo Dynamic (QCD) computation domain of about several nucleons length. Quark-gluon interaction is just beginning within a proton-occupied region of empty space.

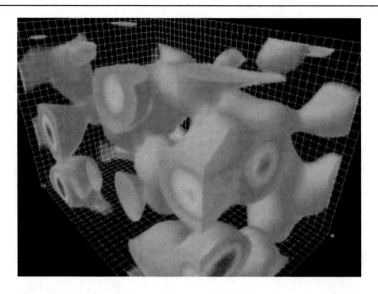

Figure 8. Snapshot of mass-less energy structures "gluon fields" quickly forming out of QCD vacuum to swiftly re-arrange and push 3 quarks about inside a proton. Gluon fields appear-expand-twist-shrink-twist-disappear into vacuum trillions of times a second.

Each quark in each proton in each of the millions of cells within humans is completely- vigorously-energetically re-arranged (renewed) trillions of times per second by the QCD quantum vacuum – to give incredibly-long healthy proton life. And long healthy proton life is vital for us all. For each proton and its electric charge are vital parts of complex energy generation/conversion processes in human cells. So, even if the vacuum never gives us energy for space flight, I am grateful for its better gift: energy for maintaining all our human life.

2.3. ENERGY AND THRUST FOR FIELD PROPULSION: FROM THE VACUUM OF SPACE

As previously mentioned, I was introduced to the idea of the quantum vacuum: the "zero-point" vacuum fields that underlie the ground state of the matter and radiation fields that define the nature of physical forces and things. Wheeler viewed this quantum vacuum as stupendously energetic over the shortest times and distance allowed by quantum mechanics (about 10^{-35}m and 10^{-43}s) and he claimed, as is symbolized below, this energetic action was happening continually and everywhere throughout all the entire vastness of cosmic space (Figure 9, Figure 10).

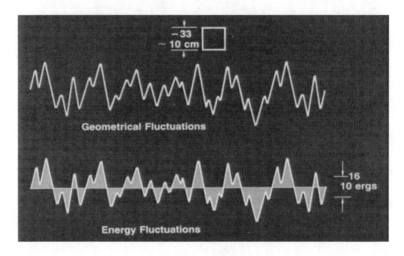

Figure 9. Time history of expectation energy (10 trillion joules) in smallest region of physics.

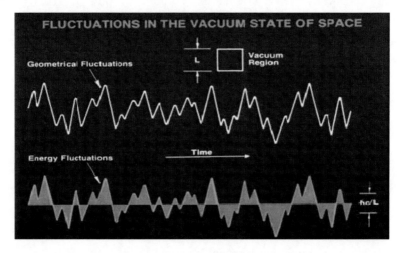

Figure 10. Expectation value of vacuum fluctuation energy inside a vacuum region of given size.

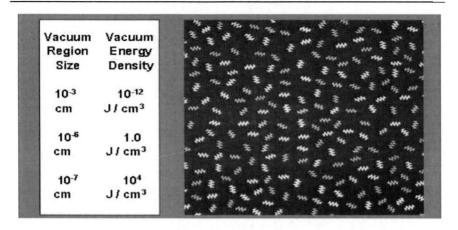

Figure 11. Fluctuation energy densities become significant at nanometer scales of distance in space.

Though briefly mentioning the Quantum Chromodynamics Vacuum, I mainly considered the electromagnetic (EM) quantum vacuum for what I called a "quantum interstellar ramjet". This vacuum is associated with electromagnetic radiation and with attraction-repulsion in atoms and molecules. It soon became evident that interacting with quantum vacuum over molecular (10^{-6} to 10^{-8} cm) scales of distance, might not be possible without being within regions where atomic forces within atomic regions might prevent vacuum energy-extracting interactions.

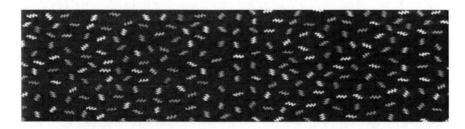

Figure 12. Small part of long total swath of zero-point energies (about 10^{30} coulombs) extracted by ramjet-like ship at 10^{-6} cm scale-of-interaction distance) during a 2.2 million light-year trip to M-31. And, during this long trip an equivalent of 3,000 tons of mass-energy is collected by the ship. About 10 billion coulombs off energy are being collected inside the ship each second at a speed of almost c, generating about 300,000 terra-watts of power during each second of such flight.

A very preliminary sizing of a quantum interstellar ramjet was made. It had roughly the same diameter, volume and length as the Apollo-Saturn 5 moon rocket. The ramjet was configured for a capture area of 100 m² and a length of 100 m. Shown below is a very short segment of the 100 square meter swath of zero-point vacuum energies swept out by this ram-jet-like ship during its very long 2.2 million light-year flight distance to the M-31 nebula (Figure 11, Figure 12).

In the 1980s Dr. Anders Hansson of the International Academy of Astronautics (IAA) read my first quantum ramjet paper and proposed that my craft might exploit the "Higgs Quantum Vacuum" associated with the Weak Nuclear force. This vacuum exists over quark scales of distance within atomic nuclei and is involved in neutrino absorption and emission; and radioactive beta decay in neutrons. So Hansson would extract energy from the Higgs vacuum rather than from Quantum-Electro-Dynamic (QED) vacuum or from the Quantum-Chromo-Dynamic (QCD) vacuum. Hansson suggested zero-point energy extraction from the Higgs quantum vacuum by the mediating action of the 3 "vector bosons" (**W⁺, W⁻** and **Z** particles). That mediate the Weak Force and are believed to have been given their rest mass by the Higgs Mechanism of the Higgs Field. Hansson outlined this idea in papers (IAA-89-667; IAA-89-669) given at 1987 and 1989 International Astronautic Federation Congresses.

The Weak force acts over only a 10^{-15} cm. scale of distance in changing up-quarks in neutrons into down-quarks in protons. So, Higgs vacuum energy densities in these small regions might be higher than my QED vacuum densities over nanometer distances. But higher energy densities within very small distance, in Higgs quantum vacuum may be very difficult to achieve. But since the Weak force participates in both nuclear fission and nuclear fusion reactions, one might consider studies of energy extraction from Higgs vacuum fields be done in conjunction with nuclear energy extraction from fission and fusion reactions.

Interestingly, we found from the work of Barrett (to be described in future chapters) that mathematical structure of classical electromagnetic radiation fields can be conditioned into the same SU(2) Lie Symmetry as the SU(2) matter fields associated with the Weak Force and Higgs quantum fields. Moreover, Barrett has noted possibility for describing the mediating action associated with the 3 vector bosons of SU(2) Weak-Force matter-field theory with the mediating action between: electric (E) fields; magnetic (B) fields and A-vector potential (A) fields of SU(2) radiation field theory. Major differences are vastly different geometric size in boundary conditions and vastly different wave frequencies of associated SU(2) radiation fields and SU(2) matter fields. In this respect, I hoped that some kind of resonance might be possible between SU(2) EM and SU(2) matter wave frequencies that would be enormous multiples of each other. Unfortunately, there seems no experimental or theoretical basis for this hope. On the other hand, nuclear fission reactions achieve beyond-breakeven, energy-amplifying, chain-reactions. And the Weak interaction (and presumably its associated Higgs quantum vacuum field) participates in such nuclear fission reactions.

It is interesting that the more robust action of the QCD vacuum and the more subtle action of the Higgs vacuum create all our universe's seen matter, with the more intense material motion generated by gluon field materialization in quark-occupied regions of empty space creating about 97 percent of the seeming solidity and substance of seen matter; Meanwhile, the mysterious Higgs vacuum creates the remaining 3 percent of seen matter's material mass - point-like material particles like quarks, electrons, muons and neutrinos. Unfortunately, the invisible quantum vacuum, just like Einstein's invisible space-time metric remains to be fully understood by science. So, enormous in both ideas is surely needed for truly effective field propulsion that could be subject to deep, detailed scientific scrutiny and engineering designs.

Figure 13 shows a detailed simulation of the intensely, energizing action off the QCD vacuum, developed by Derek Leinweber at Adelaide University. It shows in depth the evolution of the twisting-expanding-compressing gluon energy structures, which swiftly push-whirl-rotate tiny trios of quarks throughout individual proton and neutron-occupied regions of QCD vacuum. But now simulations are limited economically to several hadrons, not the thousands that participate in actual fusion reactions. An example below is a promising "clean" fusion reaction involving many hadron fusions and interactions which can be calculated by high-fidelity

Leinweber-like simulations that can reveal detail for clean, efficient reactor designs (Figure 14).

Figure 13. Dynamics of Gluon energy structures in a proton-occupied region of the QCD Vacuum.

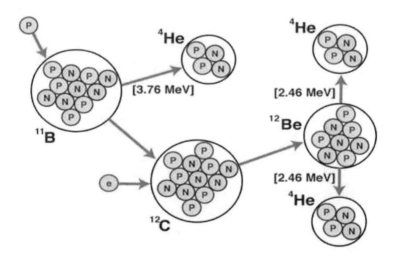

Figure 14. Favorable, clean, complex nuclear fusion reactions that need comprehensive analysis and understanding of what's taking place in proton and neutron - occupied regions of QCD vacuum.

Energy for Spaceflight Power and Propulsion ... 51

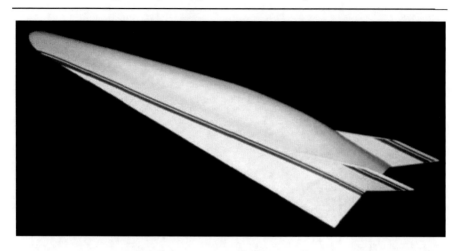

Figure 15. Field and ramjet-propelled aero-space plane that would revolutionize Earth-to-orbit travel by extracting energy and power from the atmosphere of Earth and quantum vacuum of space.

Figure 15 shows a field-propelled, ramjet-like craft that could possibly extract zero-point energy from the QED or QCD or Higgs Quantum Vacuum for almost propellant-less power and propulsion. It is an aero-space craft with "wave-rider" under-surface that emits specially-conditioned EM energy into oncoming air and quantum vacuum. Also Figure 16 is vacuum-disturbing EM radiation from the vehicle under-surface causing vacuum energy release and air heating from vacuum processes such as: virtual particle annihilation in QED vacuum; gluon-materialization in QCD vacuum; or beta particle creation in Higgs quantum vacuum.

Haisch, Rueda and Puthoff propose that an object's inertia, at least in part, is caused by the electromagnetic resistance of the quantum vacuum. Therefore, in exploring vacuum energy extraction and the vacuum's resistance to accelerated flight, we approximated the ambient zero-point vacuum fields surrounding accelerating ships by aerodynamic fields that are used in computational fluid dynamics (CFD). Figure 17 shows similar vibratory-energetic nature of recoiling-colliding air molecules in air and electromagnetic energy fluctuations in space.

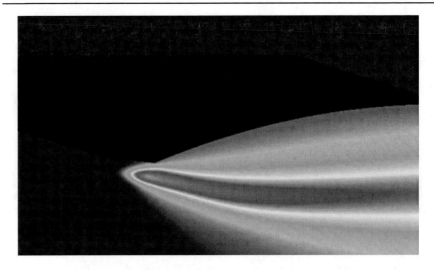

Figure 16. Favorable disturbing of incoming air and quantum vacuum by EM radiation from forward undersurface and from external air combustion, also shown is vacuum energy released into air or space at high speed.

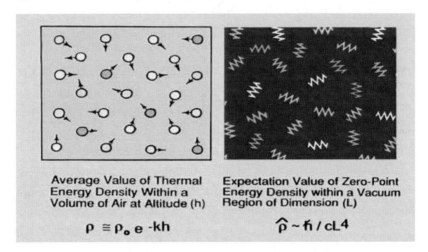

Figure 17. Energetics of Atmospheric Air and Quantum Vacuum.

Figure 18 is similarity in energy storage and dissipation in the atmosphere of Earth and the vacuum of space. This similarity allows relating ship aerodynamic and quantum dynamic behavior in air and space for given ship to sound-speed ratio and ship to light-speed- ratio.

Energy for Spaceflight Power and Propulsion ...

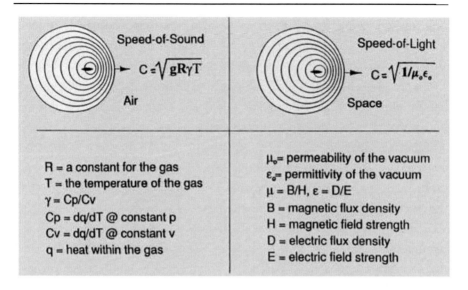

Figure 18. Propagating speed of sound and light disturbances in air and vacuum.

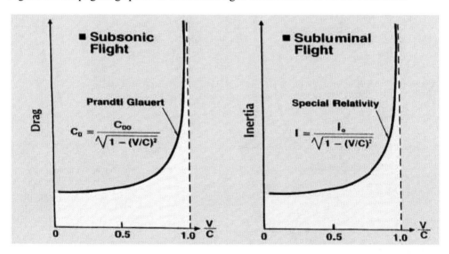

Figure 19. Flight resistance increase as ship-speed V approaches disturbance-speed c.

Just as Einstein's Special relativity predicts inertia of accelerating bodies to increase towards infinity as their speed approached propagating speed-of-light, so air resistance of accelerating bodies, (as estimated by an early aerodynamic theory) was predicted to increase to infinity as aircraft speed approached the propagating speed of sound. But this early theory assumed non- compressibility of air, which was found not true. Correct theory then revealed significant air resistance increase at sound-speed, but infinite air resistance was never found to occur (Figure 19).

Figure 20 below compression-expansion regions formed about supersonic aircraft in expandable-compressible air. General Relativity allows similar compression and expansion regions.

Einstein Special Relativity, seemingly valid for tiny point-like fundamental particles, doesn't allow faster-than-light travel. No known particle has ever been pushed beyond light-speed by the most powerful particle accelerator. But General Relativity's compressible-expandable metric may allow larger macroscopic objects to exceed light-speed, just as compressible-expandable airflow, in this CFD computation below, allows a wing-like craft to reach 90 percent; 99 percent of sound-speed – when maximum flight resistance is actually reached (Figure 21).

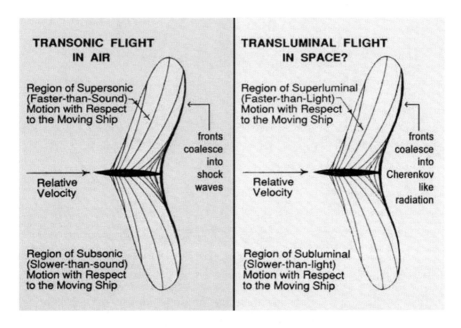

Figure 20. Compression-expansion in air and quantum vacuum at almost sound and light-speed.

Energy for Spaceflight Power and Propulsion ... 55

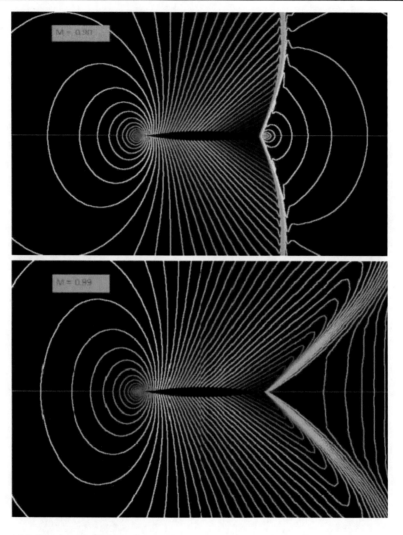

Figure 21. Maximum flight resistance may even occur at slower than the speed-of-light. Here, the maximum resistance to this moving wing is occurring at 99 percent of the speed-of-sound.

Similarity in energetic-compression-expansion of air and vacuum enabled preliminary simulation of interactions between moving ships and quantum vacuum. But difference in pressures exerted by gaseous air and mass-less vacuum prevented better simulation. Figure 22 shows inward-compressing positive pressure exerted by air on a ship, and the "negative pressure" exerted by vacuum viewed simplistically as acting in opposite, outward expanding direction.

There is a profound difference in the "positive" inwardly-compressing pressures exerted by gaseous air or EM radiation on accelerating wing-like ships and negative" outwardly- expanding pressures exerted by quantum vacuum on a similarly accelerating wing-like shape. And this is symbolized below as inwardly and outwardly acting pressures acting on moving shapes. But this results in motion through gaseous air resulting in net resistance to a moving shape, while movement through mass-less quantum vacuum results in the shape's impulsion.

But this is not observed in experiments or predicted by theory. So, such an impulsive pressure could not be from direct interaction of a ship's substance with quantum vacuum. Rather, it might be from a more complex interaction between negative pressure vacuum and positive pressure radiation that is emitted from the positive pressure substance of an accelerating ship.

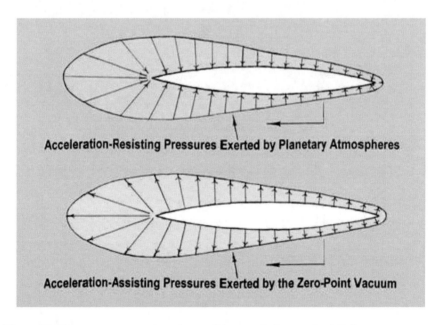

Figure 22. Vacuum pressure viewed as acting oppositely to action of air pressure on a moving shape.

A final quantum vacuum uncertainty is the largely un-addressed problem of vacuum coherency (the vacuum remaining stable long enough for its energy structures to be disturbed and interacted with for energy-extraction, reduced resistance or thrust. A similar problem is encountered in low-drag aircraft wing design, where body and wing contours must be carefully shaped to form favourable pressure gradients to maximize the distance high-lift and low-drag laminar airflow will travel over the wing and extend the time that turbulence can be delayed.

REFERENCES

[1] Sanger, E. "The attainability of the Stars", *7th International Congress*, Rome, Italy, 1956.

[2] Bussard, R.W., "Galactic Matter and Interstellar Flight", *Astronatical Acta,* Vol.6, 1960.

[3] Wheeler, J.A., "Superspace and the Nature of Quantum Geometrodynamics", *Topics in Non-linear Physics,* pp. 615-654; Proceeding of the Physics Session of the International School of Nonlinear Mathematics and Physics, Springer Verlag, 1968.

[4] Froning, H.D., Jr., "Propulsion Requirements for a Quantum Interstellar Ramjet", *Journal of the British Interplanetary Society,* Volume 36, Number 7, July, 1980.

[5] Leinweber, Derek B. *<Derek B. Leinweber, Visual QCD Archive>* Google, 2017.

[6] Hannson, P.A., "On the Use of Vacuum for Interstellar Travel", *37th International Astronautical Federation, IAA* 87-631, (1987).

[7] Hannsson, P.A., "On the Use of Vacuum for Interstellar Travel", *39th International Astronautical Federation, IAA*, 80-667, (1989).

[8] Froning, H.D. Jr., "Quantum Vacuum Engineering for Power and Propulsion from the Energetics of Space", *Proceedings of COFE 3, The Third Conference on Future Energy*, Integrity Research Institute, IntegrityResearchInstitute.org.

Chapter 3

THE EVOLUTION OF FIELD PROPULSION FROM MUCH SLOWER- THAN-LIGHT TO MUCH FASTER-THAN-LIGHT SPEED

Herman D. Froning, Jr

Field Propulsion develops force or acceleration by actions and reactions of fields, not by combustion and expulsion of matter. And it may be imperative that such propellant-less propulsion eventually, completely replace matter-emitting propulsion if global warming from carbon emissions be kept acceptably low. Nearly propellant-less field propulsion may also be imperative to truly explore our solar system and extend human presence to the stars. And actions of fields already reduce propellant in the spacecraft thruster shown below (Figure 1).

Actions and reactions of fields are also vital for needed space nuclear fusion power and propulsion systems. For such systems require strong electric or magnetic fields to overcome strong coulomb repulsion between ions at the very short distances needed for their fusion. Figure 2 shows the action of strong ion-confining magnetic fields causing ions to fuse together in a fusion reactor (a "dense mass focus" device) where electric discharges transforms flowing nuclear fuel into plasma (ions) which are being compressed very close to each other by strong magnetic pressures of hundreds of teslas until their fusion occurs.

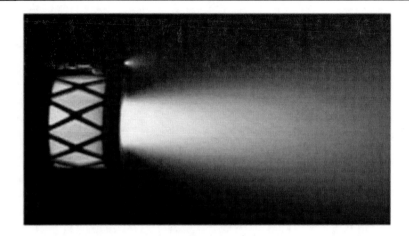

Figure 1. Spacecraft thruster.

Fusion power systems enable jet propulsion systems that consume much less propellant. For example, Flight Unlimited found that higher rocket exhaust velocities could be achieved from hotter heating of hydrogen propellant by the greater energies released from nuclear fusion, compared to less energy release and lower exhaust velocities from chemical combustion of hydrogen and oxygen. This enabled a 4-5 fold propellant reduction to achieve needed thrust and a 5-8 fold reduction in the takeoff mass of interplanetary rocket ships.

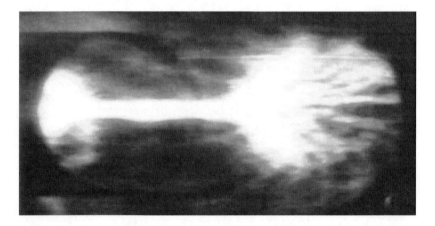

Figure 2. Action of strong ion-confining magnetic fields causing ions to fuse together in a fusion reactor.

Evolution of field propulsion might logically begin with vehicles that would be no faster than jet-propelled vehicles of today. These vehicles would generate field energy or force that was achievable with available with the science and technical understanding. During the early days of field propulsion, when field propulsion science and technology is still being developed, these field-augmented vehicles might generate thrust levels and speeds that are comparable to those of existing propulsion systems. Practical needs for early field-propelled ships might well be less vehicle cost and more vehicle safety instead of faster flight speed. Also, these ships might not rely entirely on field propulsion. Instead, they might retain some jet-propulsion engines to provide more redundancy for flight safety. However, later and more perfected field-propelled ships could travel much faster than the fastest spacecraft of today to explore all interesting celestial bodies of our solar system, and possibly even colonize some.

MUCH SLOWER THAN LIGHT FIELD PROPULSION

Initial use of the actions of-fields instead of the combustion of matter was accomplished in the design of 2 Earth-to-orbit aerospace vehicles by Flight Unlimited for the U.S. Air Force, these vehicles to revolutionize its aerospace operations soon after 2025 [1]. Both vehicles generated enormous electric power by magneto-hydro-dynamic (MHD) field actions inside 2 hypersonic jet engines- to provide vehicle air breathing propulsion to Mach 14 speed and generate about 20 MW of electrical power after about Mach 10 was reached. MHD-generated electrical power then ignited the vehicle's fusion reactor which then generated power for air breathing and for rocket propulsion until orbital speed and altitude was reached (Figures 3, 4, 5).

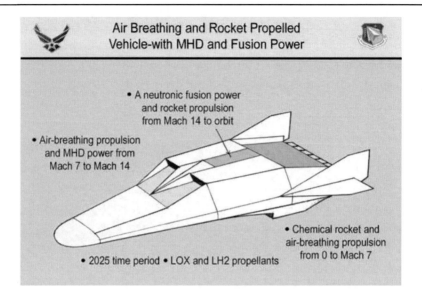

Figure 3. 250 ton reusable, single-stage-to-orbit aerospace plane - to dramatically lower Air-Force space flight costs and increase spaceflight safety by aircraft-like configuration and flight operations.

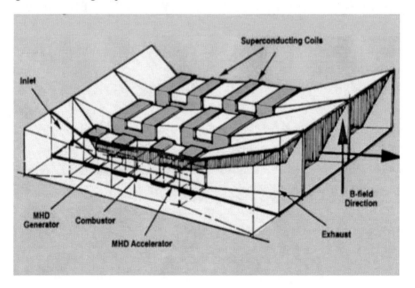

Figure 4. Air breathing propulsion/and MHD power system concept by Paul Czysz of St Louis University. This combined system eliminated very expensive vehicle ground operations by providing many megawatts of electrical power for in-flight ignition of vehicle fusion reactor at Mach 10 peed.

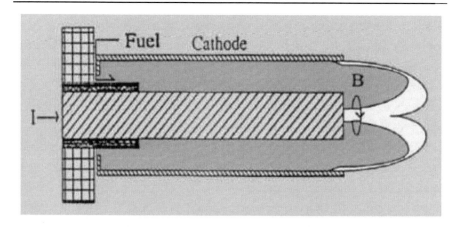

Figure 5. Fusion power and propulsion system (ignited by an air breathing/MHD system. This system provided thrust from Mach 10 to orbital speed by heating of airflow and hydrogen propellant in vehicle engine nozzle region by electron beams.

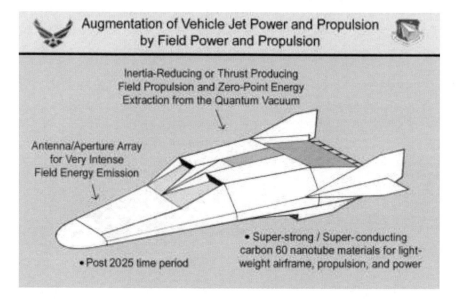

Figure 6. Vehicle jet power and propulsion by field power and propulsion.

Figure 6 shows the field and jet-propelled vehicle next configured by Flight Unlimited for the US Air Force for a later 2050 time period. This vehicle retained air breathing and rocket jet propulsion and MHD and fusion power systems and added a field propulsion module to significantly extend its capability to travel between the Earth and the Moon. Here, added propulsive velocity was needed for moon emergency rescue missions. This required ship-emitted EM fields that would be specially-conditioned to favorably interact with those ambient fields that give rise to gravity and inertia. One EM field possibility for this is the polarization-modulated EM system by T.W. Barrett [2]. It is shown in Figure 7.

Field propulsion improved flight performance and it also improved flight safety by reducing structural fatigue from repeated exposure to high pressures and temperatures during many trips between Earth and space. Figure 8 is ship EM energy emitted in fore and aft directions into its flight medium (air and space and the quantum vacuum that permeates both). This beamed energy was found to reduce resistance of both air and quantum vacuum during vehicle acceleration to high speed [3]. The EM beams reduced the vacuum's resistance by reducing its electrical permittivity, magnetic permeability and light-speed c. And they also reduced air temperatures and pressures exerted on the vehicle during each return from space.

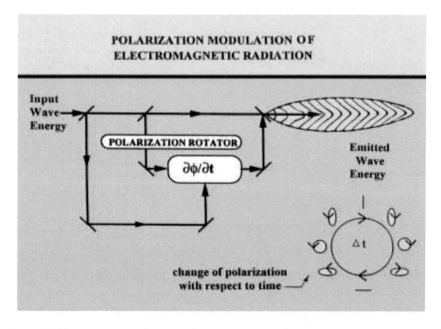

Figure 7. Polarization modulation of electromagnetic radiation.

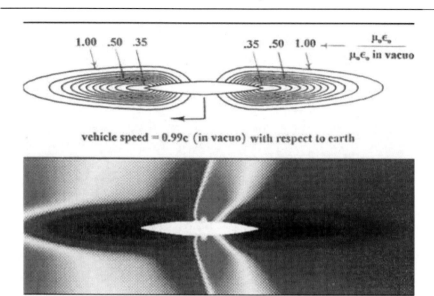

Figure 8. Pressures formed about an accelerating, EM beam-generating ship. It is experiencing impulsion (not drag) at M 0.99 speed. Darker regions are the lower air pressures within much cooler air.

The jet and field-propelled 2050 Air-Force vehicle was small and light enough to operate from existing Air- Force bases. And the vehicle's very modest propellant needs allowed very modest operational costs. So after the Air-Force Study, we configured a version of this field- propelled aerospace plane for commercial air and space flight during the 2050 time-period. A cost analysis [4] of a space-airline or space-tourism fleet of 5 vehicles indicated that they could give significant numbers of Earth's people the inspiring experience of space-flight as 144,000 passengers were flown profitably and safely between Earth and space in 12 years.

Past investigations indicated perfection of nano-technology and molecular manufacturing to a significant degree would enable continual self-replication of an initial fleet of slower-than- light starships and this initially slow, but ever-expanding and numerically-increasing fleet would eventually and inevitably colonize our entire Milky-Way galaxy over times of the order of 100,000 years. On the other hand, relativistic field propulsion that swiftly accelerated ships to nearly light-speed, could allow the settling of our galaxy to begin sooner than about 20 years. And faster-than-light field-propelled craft could allow such swift settling of our galaxy with many colonies that could conceivably be organized into communicating galactic civilizations by the faster-than-light commerce by many faster-than-light ships (Figure 9).

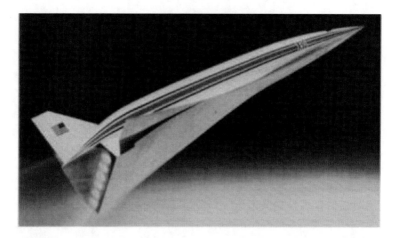

Figure 9. Solar System and Interstellar Flight.

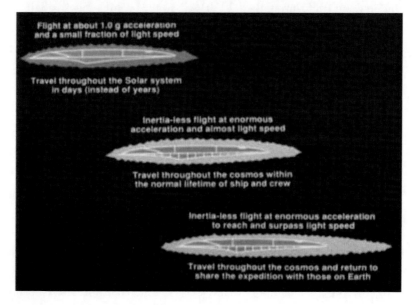

Figure 10. Evolution of Field power and propulsion: from Solar System to Cosmic exploration.

Figure 10 may be a much too "optimistic vision" of field propulsion's future. For it may not be necessarily the destiny of humanity to colonize our Cosmos by nearly propellant-less field power and field propulsion. But there is surely now need for reasonably rapid perfection of nearly- propellant-less terrestrial power and propulsion for drastic slowing of global-warming during the rest of this century. So, nearly-propellant-less field propulsion and power developed for spaceflight would surely vitally contribute to the prolonging of acceptable life on Earth.

REFERENCES

[1] Froning, H.D. Jr., Czysz, P., *"Advanced Technology and Advanced Breakthrough Propulsion Physics for 2025, 2050 Military Vehicles"*, Space Technologies and Applications International Forum, (STAIF 2006), AIP Conference Proceedings, Editor S. El Genk, AIP, Melville New York, 2006.

[2] Barrett, T.W. *"The distinction between fields and their metric"*, annals de la Louis be Broglie, 14, 1, 1989.

[3] Froning, H.D. Jr., Roach, R.L., "AIAA 2000-3478, Preliminary simulation of Vehicle Interactions with the Zero-Point Vacuum", *36th AIAA/ASME/SAE/ASEE Joint Propulsion and Exhibit*, July 17-19, Huntsville, AL, 2000.

[4] Froning, H.D. Jr., "AIAA 96-4329, Economic and Technical Challenges of Expanding Space Commerce by RLV Development", *AIAA Space Programs and Technologies Conference*, Huntsville, AL, 1996.

Chapter 4

SPACE DRIVE PROPULSION: TYPICAL FIELD PROPULSION SYSTEM

Yoshinari Minami

As previously stated in Chapter 1, field propulsion is the concept of propulsion theory of spaceship not based on usual momentum thrust but based on pressure thrust derived from an interaction of the spaceship with external fields.

In this chapter, the notion of Space Drive Propulsion is described in detail based on General Relativity (Refer to Appendix A: Outline of General Relativity). The curvature of space plays a significant role for the propulsion theory. Further, this chapter describes the behavior of space as continuum. This notion regarding the space is foundation for investigating field propulsion.

The space drive propulsion system proposed here is one of field propulsion system utilizing the action of the medium of strained or deformed field of space, and is based on the propulsion principle of the kind of pressure thrust. Figure 1 shows the basic propulsion principle of common to all kinds of field propulsion system. As shown in Figure 1, the propulsion principle of space drive propulsion system is not momentum thrust but pressure thrust induced by a pressure gradient (or potential gradient) of the space-time field (or vacuum field) between bow and stern of a spaceship. Since the pressure of the vacuum field is high in the rear vicinity of spaceship, the spaceship is pushed from the vacuum field. Pressure of vacuum field in the front vicinity of spaceship is low, so the spaceship is pulled from the vacuum field. In the front vicinity of spaceship, the pressure of vacuum field is not necessarily low but the ordinary vacuum field, that is, just as only a high pressure of vacuum field in the rear vicinity of spaceship. The spaceship is propelled by this distribution of pressure of the vacuum field. Vice versa, it is the same principle that the pressure of vacuum field in the front vicinity of spaceship is just only low and the pressure of vacuum field in the rear vicinity of spaceship is ordinary. In any case, the pressure gradient from the vacuum field (potential gradient) is formed over the entire range of the spaceship, so that the spaceship is propelled by pushing from the pressure gradient resulting from the vacuum field.

In this chapter, space drive propulsion system considered comparatively well as a field propulsion system is introduced in detail.

Propulsion Principle

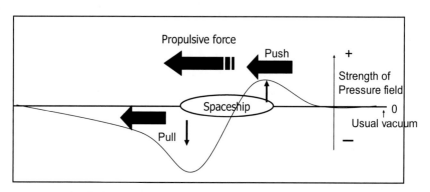

Asymmetrically interaction with the pressure of field creates propulsive force for the spaceship.

The strength of pressure field ahead of the spaceship is diminished and its behind increased, this would result in favorable pressure gradients.

Figure 1. Fundamental propulsion principle of Space Drive Propulsion.

4.1. BACKGROUND

A concept of a space drive propulsion system as a paper entitled "Space Strain Propulsion System" was introduced by Minami in 1988 [1]. The term of "space strain" was changed to "space drive" receiving the recommendation by Robert L. Forward [2]. After then, the second paper entitled "Possibility of Space Drive Propulsion" was presented at the 45th IAF in 1994 [3].

Assuming that space vacuum is an infinite continuum, the propulsion principle utilizes the pressure field derived from the geometrical structure of space, by applying both continuum mechanics and General Relativity to space. The propulsive force is a pressure thrust that arises from the interaction of space-time around the spaceship external environment and the spaceship itself; the spaceship is propelled against the space-time continuum structure. This means that space can be considered as a kind of transparent elastic field. That is, space as a vacuum performs the motions of deformation such as expansion, contraction, elongation, torsion and bending. The latest expanding universe theory (Friedmann, de Sitter, inflationary cosmological model) supports this assumption. Space can be regarded as an elastic body like rubber.

General Relativity based on Riemannian geometry implies that space is curved by the existence of energy (mass etc.). If we admit this space curvature, space is assumed as an elastic body. According to the continuum mechanics, the elastic body has the property of the motion of deformation such as expansion, contraction, elongation, torsion and bending. General Relativity uses just only the curvature of space. Expansion and contraction of space are used in Cosmology, and a theory using torsion is also studied by Hayasaka [4]. Perhaps, Twistor Theory by Roger Penrose is also applied to the torsion of space [5].

The principle of space drive propulsion system is derived from General Relativity and the theory of continuum mechanics. We assume the so-called "vacuum" of space as an infinite elastic body like rubber. The curvature of space plays a significant role for the propulsion theory. From the gravitational field equation, the strong magnetic field as well as mass density generates the curvature of space and this curved space region produces the uni-directional acceleration field. The spaceship in the curved space can be propelled in a single direction. Since the force they produce acts uniformly on every atom inside the spaceship, accelerations of any magnitude can be produced with no strain on the crews, that is, there is no action of inertial force because the thrust is a body force (i.e., it is equivalent to free-fall). Minami derived the equation of curvature of space induced by magnetic field in 1988 [1].

It was found that this equation was accordance with the equation that Levi-Civita considered (i.e., the static magnetic field creates scalar curvature) [6] by Minami in 1995.

In the latest cosmology, the terms vacuum energy and cosmological term " Λg^{ij} " are used synonymously. Λ is known as the cosmological constant. The term with the cosmological constant is identical to the stress-energy tensor associated with the vacuum energy. The properties of vacuum energy, i.e., cosmological term are crucial to expansion of the Universe, that is, to inflationary cosmology.

In the beginning, the acceleration generated by curvature of space induced by strong magnetic field based on external and internal Schwarzschild solution was studied [3]. However, superior acceleration based on de Sitter solution is obtained and presented at 47th IAF in 1996 [7]. The details are published in JBIS, Vol.50 [8] and presented at STAIF-98 in 1998 [9].

At the present day, space drive propulsion system is based on the de Sitter solution as follows:

Basically: *The acceleration derived from the de Sitter solution does not require a strong magnetic field. At the present day, space drive propulsion system based on the de Sitter solution needs not strong magnetic field but the technology to excite space.*

Additionally, inflationary universe which shows the rapid expansion of space is based on the phase transition of the vacuum exhibited by the Weinberg-Salam model of the electroweak interaction. The vacuum as space has the property of a phase transition, just like water may become ice and vice versa. This shows that vacuum as space possesses a substantial physical structure such as material. It coincides with the precondition of a space drive propulsion principle. As is well known in cosmology, the expansion rule of the universe is governed by the Friedman's equations and the Robertson-Walker metric.

In the last section, the propulsion principle for this space drive is newly studied from another angle, that is, the pressure of the field induced by local rapid expansion of space is completely considered in the propulsion principle based on the latest cosmology.

The acceleration performance of this system is found by the solution of the gravitational field equation, such as the Schwarzschild solution, Reissner-Nordstrom solution, Kerr solution, and de Sitter solution. The details of their solutions are described in Appendix A.

4.2. FUNDAMENTAL MECHANICAL STRUCTURE OF SPACE

Based on General Relativity, this section describes the behavior of space. This notion regarding the space is basic foundation for investigating field propulsion.

Fundamental Supposition of Space

On the supposition that space is an infinite continuum, continuum mechanics can be applied to the so-called "vacuum" of space. This means that space can be considered as a kind of transparent field with elastic properties. Figure 2 shows the curvature of space.

If space curves, then an inward normal stress "–P" is generated (Figure 2(a)). This normal stress, i.e., surface force serves as a sort of pressure field.

$$-P = N \cdot (2R^{00})^{1/2} = N \cdot (1/R_1 + 1/R_2), \tag{1}$$

where N is the line stress, R_1, R_2 are the radius of principal curvature of curved surface, and R^{00} is the major component of spatial curvature.

A large number of curved thin layers form the unidirectional surface force, i.e., acceleration field. Accordingly, the spatial curvature R^{00} produces the acceleration field α (Figure 2(b)).

The fundamental three-dimensional space structure is determined by quadratic surface structure. Therefore, a Gaussian curvature K in two-dimensional Riemann space is significant. The relationship between K and the major component of spatial curvature R^{00} is given by:

$$K = \frac{R_{1212}}{(g_{11}g_{22} - g_{12}^{2})} = \frac{1}{2} \cdot R^{00}, \qquad (2)$$

where R_{1212} is non-zero component of Riemann curvature tensor.

It is now understood that the membrane force on the curved surface and each principal curvature generates the normal stress "–P" with its direction normal to the curved surface as a surface force. The normal stress "–P" acts towards the inside of the surface as shown in Figure 2 (a).

A thin-layer of curved surface will take into consideration within a spherical space having a radius of R and the principal radii of curvature that are equal to the radius ($R_1 = R_2 = R$). Since the membrane force N (serving as the line stress) can be assumed to have a constant value, Eq. (1) indicates that the curvature R^{00} generates the inward normal stress "–P" of the curved surface. The inwardly directed normal stress serves as a pressure field.

When the curved surfaces are included in a great number, some type of unidirectional pressure field is formed. A region of curved space is made of a large number of curved surfaces and they form the field as a unidirectional surface force (i.e., normal stress). Since the field of the surface force is the field of a kind of force, the force accelerates matter in the field, i.e., we can regard the field of the surface force as the acceleration field. A large number of curved thin layers form the unidirectional acceleration field (Figure 2b)). Accordingly, the spatial curvature R^{00} produces the acceleration field α.

Therefore, the curvature of space plays a significant role to generate pressure field.

Applying membrane theory, the following equilibrium conditions are obtained in quadratic surface,

$$N^{\alpha\beta}b_{\alpha\beta} + P = 0, \qquad (3)$$

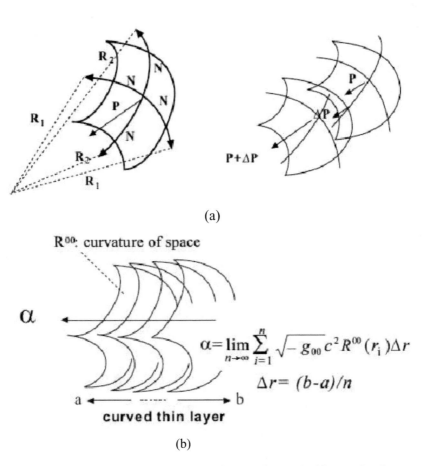

Figure 2. Curvature of Space: (a) curvature of space plays a significant role. If space curves, then inward stress (surface force) "P" is generated ⟹ A sort of pressure field; (b) a large number of curved thin layers form the unidirectional surface force, i.e., acceleration field α.

where $N^{\alpha\beta}$ is a membrane force, i.e., line stress of curved space, $b_{\alpha\beta}$ is second fundamental metric of curved surface, and P is the normal stress on curved surface [10, 11].

The second fundamental metric of curved space $b_{\alpha\beta}$ and principal curvature $K_{(i)}$ has the following relationship using the metric tensor $g_{\alpha\beta}$,

$$b_{\alpha\beta} = K_{(i)} g_{\alpha\beta}. \tag{4}$$

Therefore we get:

$$N^{\alpha\beta} b_{\alpha\beta} = N^{\alpha\beta} K_{(i)} g_{\alpha\beta} = g_{\alpha\beta} N^{\alpha\beta} K_{(i)} = N_\alpha^{\ \alpha} K_{(i)} = N \cdot K_{(i)}. \tag{5}$$

From Eq. (3) and Eq. (5), we get:

$$N_\alpha^{\ \alpha} K_{(i)} = -P. \tag{6}$$

As for the quadratic surface, the indices α and i take two different values, i.e., 1 and 2, therefore Eq. (6) becomes:

$$N_1^{\ 1} K_{(1)} + N_2^{\ 2} K_{(2)} = -P, \tag{7}$$

where $K_{(1)}$ and $K_{(2)}$ are principal curvature of curved surface and are inverse number of radius of principal curvature (i.e., *1/R₁* and *1/R₂*).

The Gaussian curvature K is represented as:

$$K = K_{(1)} \cdot K_{(2)} = (1/R_1) \cdot (1/R_2). \tag{8}$$

Accordingly, suppose $N_1^{\ 1} = N_2^{\ 2} = N$, we get:

$$N \cdot (1/R_1 + 1/R_2) = -P. \tag{9}$$

It is now understood that the membrane force on the curved surface and each principal curvature generate the normal stress "–P" with its direction normal to the curved surface as a surface force. The normal stress "–P" is towards the inside of surface as shown in Figure 2.

A thin-layer of curved surface will be taken into consideration within a spherical space having a radius of R and the principal radii of curvature which are equal to the radius ($R_1 = R_2 = R$). From Eqs. (2) and (8), we then get:

$$K = \frac{1}{R_1} \cdot \frac{1}{R_2} = \frac{1}{R^2} = \frac{R^{00}}{2}. \tag{10}$$

Considering $N \cdot (2/R) = -P$ of Eq. (9), and substituting Eq. (10) into Eq. (9), the following equation is obtained:

$$-P = N \cdot \sqrt{2R^{00}}. \tag{11}$$

Since the membrane force N (serving as the line stress) can be assumed to have a constant value, Eq. (11) indicates that the curvature R^{00} generates the inward normal stress "–P" of the curved surface. The inwardly directed normal stress serves as a kind of pressure field.

Here, we give an account of curvature R^{00} in advance. The solution of metric tensor $g^{\mu\nu}$ is found by gravitational field equation as the following:

$$R^{\mu\nu} - \frac{1}{2} \cdot g^{\mu\nu} R = -\frac{8\pi G}{c^4} \cdot T^{\mu\nu}, \tag{12}$$

where $R^{\mu\nu}$ is the Ricci tensor, R is the scalar curvature, G is the gravitational constant, c is the velocity of light, $T^{\mu\nu}$ is the energy momentum tensor.

Furthermore, we have the following relation for scalar curvature R :

$$R = R^\alpha{}_\alpha = g^{\alpha\beta} R_{\alpha\beta}, \quad R^{\mu\nu} = g^{\mu\alpha} g^{\nu\beta} R_{\alpha\beta}, \quad R_{\alpha\beta} = R^j{}_{\alpha j\beta} = g^{ij} R_{i\alpha j\beta}. \quad (13)$$

Ricci tensor $R^{\mu\nu}$ is represented by:

$$R_{\mu\nu} = \Gamma^\alpha_{\mu\alpha,\nu} - \Gamma^\alpha_{\mu\nu,\alpha} - \Gamma^\alpha_{\mu\nu}\Gamma^\beta_{\alpha\beta} + \Gamma^\alpha_{\mu\beta}\Gamma^\beta_{\nu\alpha} \quad (= R_{\nu\mu}), \quad (14)$$

where $\Gamma^i{}_{jk}$ is Riemannian connection coefficient.

If the curvature of space is very small, the term of higher order than the second can be neglected, and Ricci tensor becomes:

$$R_{\mu\nu} = \Gamma^\alpha_{\mu\alpha,\nu} - \Gamma^\alpha_{\mu\nu,\alpha}. \quad (15)$$

The major curvature of Ricci tensor ($\mu = \nu = 0$) is calculated as follows:

$$R^{00} = g^{00} g^{00} R_{00} = -1 \times -1 \times R_{00} = R_{00}. \quad (16)$$

As previously mentioned, Riemannian geometry is geometry that deals with curved Riemann space, therefore, Riemann curvature tensor is the principal quantity. All components of Riemann curvature tensor are zero for flat space (strictly speaking, only 20 independent components of Riemann curvature tensor R_{pijk} are zero) and non-zero for curved space. If an only non-zero component of Riemann curvature tensor exists, the space is not flat space but curved space. Although Ricci tensor $R^{\mu\nu}$ has 10 independent components, the major component is the case of $\mu = \nu = 0$, i.e., R^{00}. Therefore, the curvature of space plays a significant role.

Other Standpoint of View of Space (Fundamental Structure Equation of Space)

The mechanical structure due to space strain is described here from another point of view.

As shown in Figure 3, if the line element between the arbitrary two near points (A and B) in space region **S** (before structural deformation) is defined as $ds = g_i dx^i$, the infinitesimal distance between the two near points is given by

$$ds^2 = g_{ij} dx^i dx^j. \tag{17}$$

Let us assume that a space region **S** is structurally deformed by external physical action and transformed to space region **T**. In the deformed space region **T**, the line element between the identical two near point (A' and B') of the identical space region newly changes, differs from the length and direction, and becomes $ds' = g'_i dx^i$. Therefore, the infinitesimal distance between the two near points using the convected coordinate ($x'^i = x^i$) is given by

$$ds'^2 = g'_{ij} dx^i dx^j. \tag{18}$$

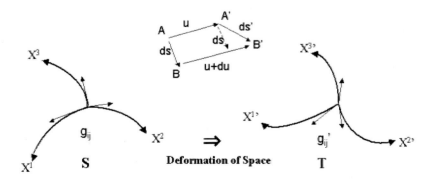

Figure 3. Fundamental structure of Space.

The g'_i is the transformed base vector from the original base vector g^i and the g'_{ij} is the transformed metric tensor from the original metric tensor g_{ij}. Since the degree of deformation can be expressed as the change of distance between the two points, we get

$$ds'^2 - ds^2 = g'_{ij} dx^i dx^j - g_{ij} dx^i dx^j = (g'_{ij} - g_{ij}) dx^i dx^j = r_{ij} dx^i dx^j.$$
(19)

Hence the degree of geometrical and structural deformation can be expressed by the quantity denoted change of metric tensor, i.e.,

$$r_{ij} = g'_{ij} - g_{ij}.$$
(20)

On the other hand, the state of deformation can be also expressed by the displacement vector "u".

From the continuum mechanics [10, 11, 12], we use the following equations;

$$du = g^i u_{i:j} dx^j,$$
(21)

$$ds' = ds + du = ds + g^i u_{i:j} dx^j.$$
(22)

Here we use the usual notation ":" for covariant differentiation. As is well known, the partial derivative $u_{i,j} = \dfrac{\partial u_i}{\partial x^j}$ is not tensor equation. The covariant derivative $u_{i:j} = u_{i,j} - u_k \Gamma^k_{ij}$ is tensor equation and can be carried over into all coordinate systems.

From the usual continuum mechanics, the infinitesimal distance after deformation becomes (see Appendix B)

$$ds'^2 - ds^2 = r_{ij}dx^i dx^j = (u_{i:j} + u_{j:i} + u^k{}_{:i}u_{k:j})dx^i dx^j. \tag{23}$$

The terms of higher order than second $u^k{}_{:i}u_{k:j}$ can be neglected if the displacement is enough small value. As the actual physical space can be dealt with the minute displacement from the trial calculation of strain, we get

$$r_{ij} = u_{i:j} + u_{j:i}. \tag{24}$$

Whereas, according to the continuum mechanics, the strain tensor e_{ij} is given by

$$e_{ij} = \frac{1}{2} \cdot r_{ij} = \frac{1}{2} \cdot (u_{i:j} + u_{j:i}). \tag{25}$$

So, we get

$$ds'^2 - ds^2 = (g'_{ij} - g_{ij})dx^i dx^j = 2e_{ij}dx^i dx^j, \tag{26}$$

where g'_{ij}, g_{ij} is a metric tensor, e_{ij} is a strain tensor, and $ds'^2 - ds^2$ is the square of the infinitesimal distance between two infinitely proximate points x^i and $x^i + dx^i$.

Eq. (26) indicates that a certain geometrical structural deformation of space is shown by the concept of strain. In a word, the change of metric tensor $(g'_{ij} - g_{ij})$ due to the existence of mass energy or electromagnetic energy tensor produces the strain field e_{ij}.

Eq. (26) indicates that a certain geometrical structural deformation of space is described by the concept of strain. Since the space-time is distorted, the infinitesimal distance between two infinitely proximate points x^i and x^i + dx^i is important for geometry of the space-time; the physical strain is generated by the difference of geometrical metric of space-time. Namely, a certain structural deformation is described by strain tensor e_{ij}. From Eq. (26), the strain of space is described as follows:

$$e_{ij} = 1/2 \cdot (g_{ij}' - g_{ij}). \tag{27}$$

Eq. (27) indicates that a certain geometrical structural deformation of space is shown by the concept of strain. In essence, the change of metric tensor $(g_{ij}' - g_{ij})$ due to the existence of mass energy or electromagnetic energy tensor produces the strain field e_{ij}. Namely, a certain structural deformation of space-time is described by strain tensor e_{ij}; the physical strain is generated by the difference of a geometrical metric of space-time.

Mechanics of Space

As is well known in the continuum mechanics, the elastic force F^i is given by the gradient of stress tensor σ^{ij}, and using elastic law with the elastic modulus $E^{ij\mu\nu}$, if we apply the continuum mechanics to above result, the strain tensor e_{ij} produces the stress field σ^{ij} and from the equilibrium conditions of continuum, we get

$$-F^i = \sigma^{ij}{}_{:j} \text{ and } \sigma^{ij} = E^{ij\mu\nu} e_{\mu\nu}. \tag{28}$$

The stress tensor σ^{ij} is a surface force and F^i is a body force. The body force is an equivalent gravitational action because of acting all elements of space uniformly. From Eq. (28), the following equation is obtained

$$-F^i = \sigma^{ij}{}_{:j} = (E^{ij\mu\nu} e_{\mu\nu})_{:j} = E^{ij\mu\nu} e_{\mu\nu:j}. \tag{29}$$

Here, we assume that $E^{ij\mu\nu}$ is constant for covariant differentiation.

Space Drive Propulsion

Furthermore, expanding the concept of vector parallel displacement in Riemann space, the following equation has newly been obtained (see Appendix B):

$$\omega_{\mu\nu} = R_{\mu\nu kl}\, dA^{kl}, \tag{30}$$

where $\omega_{\mu\nu}$ is rotation tensor, dA^{kl} is infinitesimal areal element.

According to the nature of Riemann curvature tensor $R_{\mu\nu kl}$, $\omega_{\mu\nu}$ indicates the rotation of displacement field. Eq. (30) indicates that a curved space produces the rotation of displacement field in the region of space. Now, the rotation tensor $\omega_{\mu\nu}$ and strain tensor e_{ij} satisfy the following differential equation in continuum mechanics:

$$\omega_{\mu\nu,j} = e_{\nu j,\mu} - e_{\mu j,\nu}. \tag{31}$$

This equation is true on condition that the order of differential can be exchanged in a flat space. To expand above equation into a curved Riemann space, the equation shall be transformed to covariant differentiation and it is possible on condition of $\Gamma^{\alpha}_{j\nu} e_{\mu\alpha} = \Gamma^{\alpha}_{j\mu} e_{\nu\alpha}$.

Thus, we obtain

$$\omega_{\mu\nu.j} = e_{\nu j:\mu} - e_{\mu j:\nu}. \tag{32}$$

Here we use the usual notation ":" for covariant differentiation. As is well known, the partial derivative $u_{i,j} = \dfrac{\partial u_i}{\partial x^j}$ is not tensor equation. The covariant derivative $u_{i:j} = u_{i,j} - u_k \Gamma^{k}_{ij}$ is tensor equation and can be carried over into all coordinate systems.

Eq. (32) indicates that the displacement gradient of rotation tensor corresponds to difference of the displacement gradient of strain tensor.

Here, if we multiply both sides of Eq. (32) by fourth order tensor denoted the nature of space $E^{ij\mu\nu}$ formally, we obtain

$$E^{ij\mu\nu}\omega_{\mu\nu.j} = E^{ij\mu\nu}(R_{\mu\nu kl}dA^{kl})_{:j} = E^{ij\mu\nu}R_{\mu\nu kl:j}dA^{kl} \qquad (33)$$

and

$$E^{ij\mu\nu}e_{\nu j:\mu} - E^{ij\mu\nu}e_{\mu j:\nu} =$$
$$(E^{ij\mu\nu}e_{\nu j})_{:\mu} - (E^{ij\mu\nu}e_{\mu j})_{:\nu} = \sigma^{i\mu}_{\;:\mu} - \sigma^{i\nu}_{\;:\nu} = \Delta\sigma^{ir}_{\;:r} \qquad (34)$$

As is well known in the continuum mechanics [10, 11], the relationship between stress tensor σ_{ij} and strain tensor e_{ml} is given by

$$\sigma^{ij} = E^{ijml}e_{ml}. \qquad (35)$$

Furthermore, the relationship between body force F^i and stress tensor σ_{ij} is given by

$$F^i = \sigma^{ij}_{\;:j}, \qquad (36)$$

from the equilibrium conditions of continuum. That is, the elastic force F^i is given by the gradient of stress tensor σ_{ij}.

Therefore, Eq. (34) indicates the difference of body force ΔF^i. Accordingly, from Eqs. (33) and (34), the change of body force ΔF^i ($= \Delta\sigma^{ir}_{\;:r}$) becomes

$$\Delta F^i = E^{ij\mu\nu}R_{\mu\nu kl:j}dA^{kl}. \qquad (37)$$

Here, we assume that $E^{ij\mu\nu}$ is constant for covariant differentiation, A^{kl} is area element.

The stress tensor σ_{ij} is a surface force and F^i is a body force. The body force is an equivalent gravitational action because of acting all elements of space uniformly.

Eq. (37) indicates that the gradient of Riemann curvature tensor implying space curvature produces the body force as a space strain force. The non-zero component of Eq. (37) is just only one equation as follows:

$$F^3 = F = E^{3330}(R_{3030}A^{30})_{:3} = E^{3330} \cdot \partial(R_{3030}A^{30})/\partial r. \qquad (38)$$

As described above, an important analytical method relating the concept of continuum mechanics as a deformation with the concept of Riemannian geometry is the concept of the parallel displacement of vector.

All kinds of new researches rely on the fine-structure of space as a vacuum. It is also worth noting that this result yields the principle of constancy of light velocity in Special Relativity.

4.3. SPACE DRIVE PROPULSION THEORY

The theory of space drive propulsion is summarized as follows.

1) On the supposition that space is an infinite continuum, continuum mechanics can be applied to the so-called "vacuum" of space. This means that space can be considered as a kind of transparent elastic field. That is, space as a vacuum performs the motion of deformation such as expansion, contraction, elongation, torsion and bending. The latest expanding universe theory (Friedmann, de Sitter, inflationary cosmological model) supports this assumption. We can regard space as an infinite elastic body like rubber.
2) From General Relativity, the major component of curvature of space (hereinafter referred to as the major component of spatial curvature) R^{00} can be produced by not only mass density but also magnetic

field B as follows (see Appendix C: Curvature Control by Magnetic Field):

$$R^{00} = \frac{4\pi G}{\mu_0 c^4} B^2 = 8.2 \times 10^{-38} B^2, \qquad (39)$$

where $\mu_0 = 4\pi \times 10^{-7} (H/m)$, $c = 3 \times 10^8 (m/s)$, $G = 6.672 \times 10^{-11} (N \cdot m^2/kg^2)$, B is a magnetic field density with Tesla and R^{00} is a major component of spatial curvature ($1/m^2$).

Eq. (39) indicates that the major component of spatial curvature can be controlled by a magnetic field.

3) If space curves, then an inward normal stress "$-P$" is generated (see Figure 2).

This normal stress, i.e., surface force serves as a sort of pressure field.

$$-P = N \cdot (2R^{00})^{1/2} = N \cdot (1/R_1 + 1/R_2), \qquad (40)$$

where N is the line stress, R_1, R_2 are the radius of principal curvature of curved surface.

A large number of curved thin layers form the unidirectional surface force, i.e., acceleration field. Accordingly, the spatial curvature R^{00} produces the acceleration field α (see Figure 2).

4) From the following linear approximation scheme for the gravitational field equation 1) weak gravitational field, i.e., small curvature limit, 2) quasi-static, 3) slow-motion approximation (i.e., $v/c \ll 1$), we get the following relation between acceleration of curved space and curvature of space:

$$\alpha^i = \sqrt{-g_{00}} c^2 \int_a^b R^{00}(x^i) dx^i, \qquad (41)$$

where α^i: acceleration (m/s²), g_{00}: time component of metric tensor, a-b: range of curved space region(m), x^i: components of coordinate ($i = 0,1,2,3$), c: velocity of light, R^{00}: major component of spatial curvature.

Eq. (41) indicates that the acceleration field α^i is produced in curved space. The intensity of acceleration produced in curved space is proportional to both spatial curvature and the size of curved space. Eq. (41) is derived from as the following item (5).

5) As is well known in General Relativity, in the curved space region, a massive body causes the curvature of space-time around it, and a free particle responds by moving along a geodesic line in that space-time. The path of free particle is a geodesic line in space-time and is given by the following geodesic equation;

$$\frac{d^2 x^i}{d\tau^2} + \Gamma^i_{jk} \cdot \frac{dx^j}{d\tau} \cdot \frac{dx^k}{d\tau} = 0, \qquad (42)$$

where Γ^i_{jk} is Riemannian connection coefficient, τ is proper time, x^i is four-dimensional Riemann space, that is, three dimensional space ($x = x^1$, $y = x^2$, $z = x^3$) and one dimensional time ($w = ct = x^0$), c is the velocity of light. These four coordinate axes are denoted as x^i (i = 0, 1, 2, 3).

Proper time is the time to be measured in a clock resting for a coordinate system. We have the following relation derived from an invariant line element ds^2 between Special Relativity (flat space) and General Relativity (curved space):

$$d\tau = \sqrt{-g_{00}} dx^0 = \sqrt{-g_{00}} c dt \qquad (43)$$

From Eq. (42), the acceleration of free particle is obtained by

$$\alpha^i = \frac{d^2 x^i}{d\tau^2} = -\Gamma^i_{jk} \cdot \frac{dx^j}{d\tau} \cdot \frac{dx^k}{d\tau}. \qquad (44)$$

As is well known in General Relativity, in the curved space region, the massive body "m (kg)" existing in the acceleration field is subjected to the following force F^i (N):

$$F^i = m\Gamma^i_{jk} \cdot \frac{dx^j}{d\tau} \cdot \frac{dx^k}{d\tau} = m\sqrt{-g_{00}} c^2 \Gamma^i_{jk} u^j u^k = m\alpha^i, \qquad (45)$$

where u^j, u^k are the four velocity, Γ^i_{jk} is the Riemannian connection coefficient, and τ is the proper time.

From Eqs. (44), (45), we obtain:

$$\alpha^i = \frac{d^2 x^i}{d\tau^2} = -\Gamma^i_{jk} \cdot \frac{dx^j}{d\tau} \cdot \frac{dx^k}{d\tau} = -\sqrt{-g_{00}} c^2 \Gamma^i_{jk} u^j u^k. \quad (46)$$

Eq. (46) yields a more simple equation from the condition of linear approximation, that is, weak-field, quasi-static, and slow motion (speed v << speed of light c: $u^0 \approx 1$):

$$\alpha^i = -\sqrt{-g_{00}} \cdot c^2 \Gamma^i_{00}. \quad (47)$$

On the other hand, the major component of spatial curvature R^{00} in the weak field is given by

$$R^{00} \approx R_{00} = R^\mu_{0\mu 0} = \partial_0 \Gamma^\mu_{0\mu} - \partial_\mu \Gamma^\mu_{00} + \Gamma^\nu_{0\mu} \Gamma^\mu_{\nu 0} - \Gamma^\nu_{00} \Gamma^\mu_{\nu\mu}. \quad (48)$$

Although Ricci tensor $R^{\mu\nu}$ has 10 independent components, the major component is the case of $\mu = \nu = 0$, i.e., R^{00}. Therefore, the major curvature of Ricci tensor R^{00} plays a significant role.

In the nearly Cartesian coordinate system, the value of $\Gamma^\mu_{\nu\rho}$ are small, so we can neglect the last two terms in Eq. (48), and using the quasi-static condition we get

$$R^{00} \approx -\partial_\mu \Gamma^\mu_{00} = -\partial_i \Gamma^i_{00}. \quad (49)$$

From Eq. (49), we get formally

$$\Gamma^i_{00} = -\int R^{00}(x^i) dx^i. \quad (50)$$

Substituting Eq. (50) into Eq. (47), we obtain

$$\alpha^i = \sqrt{-g_{00}} c^2 \int_a^b R^{00}(x^i) dx^i. \quad (51)$$

where α^i: acceleration (m/s²), g_{00}: time component of metric tensor, a-b: range of curved space region(m), x^i: components of coordinate ($i = 0,1,2,3$), c: velocity of light, R^{00}: major component of spatial curvature (1/m²).

Eq. (51) indicates that the acceleration field α^i is produced in curved space. The intensity of acceleration produced in curved space is proportional to the product of spatial curvature R^{00} and the length of curved region.

Eq. (45) yields more simple equation from above-stated linear approximation ($u^0 \approx 1$),

$$F^i = m\sqrt{-g_{00}}\, c^2 \Gamma^i_{00} u^0 u^0 = m\sqrt{-g_{00}}\, c^2 \Gamma^i_{00} =$$

$$= m\alpha^i = m\sqrt{-g_{00}}\, c^2 \int_a^b R^{00}(x^i) dx^i . \tag{52}$$

Setting $i = 3$ (i.e., direction of radius of curvature: r), we get Newton's second law:

$$F^3 = F = m\alpha = m\sqrt{-g_{00}}\, c^2 \int_a^b R^{00}(r) dr = m\sqrt{-g_{00}}\, c^2 \Gamma^3_{00} . \tag{53}$$

The acceleration (α) of curved space and its Riemannian connection coefficient (Γ^3_{00}) are given by:

$$\alpha = \sqrt{-g_{00}}\, c^2 \Gamma^3_{00}, \quad \Gamma^3_{00} = \frac{-g_{00,3}}{2 g_{33}}, \tag{54}$$

where c: velocity of light, g_{00} and g_{33}: component of metric tensor, $g_{00,3}$: $\partial g_{00}/\partial x^3 = \partial g_{00}/\partial r$. We choose the spherical coordinates "$ct = x^0$, $r = x^3$, $\theta = x^1$, $\varphi = x^2$" in space-time.

The acceleration α is represented by the equation both in the differential form and in the integral form. Practically, since the metric is usually given by the solution of gravitational field equation, the differential form has been found to be advantageous.

6) The acceleration of space drive propulsion system is based on the solutions of gravitational field equation, which is derived from Eq. (54). The concrete acceleration derived from Eq. (54) are introduced in *Appendix A*.

4.4. PROPULSION MECHANISM OF SPACE DRIVE

Since the propulsion mechanism used magnetic field in the beginning is easy to understand, we explain it using magnetic field. At present, space drive propulsion does not need the strong magnetic field under the favor of de Sitter solution. As mentioned above, the principle of space drive propulsion system is summarized in the following.

First, it is necessary for the space to be curved. Because the curvature of flat space R^{00} is zero (strictly speaking, only 20 independent components of Riemann curvature tensor R_{pijk} are zero), then the acceleration α becomes zero. Such a curved space is generated not only by mass density but also by magnetic field or electric field. In case that the intensities of the magnetic field B and the electric field E are equal, the value of $(1/2 \cdot \varepsilon_0 E^2)$ is about seventeen figures smaller than the value of $(B^2 / 2\mu_0)$. Consequently, the electric field only negligibly contributes to the spatial curvature as compared with the magnetic field. Accordingly, it is effective that the space can be curved by a magnetic field. Since the region of curved space produces the field of acceleration, the massive body existing in this acceleration field (i.e., curved space region), is moved by thrust in accordance with Newton's second law.

In view of the above described principle of propulsion, the spaceship does not move as long as the magnetic field is static. This is because an action of magnetic field to space is in equilibrium with a reaction from space. It is consequently necessary to shut off the equilibrium state in order to actually move the spaceship. As a continuum, the space has a finite strain rate, i.e., the velocity of light. When the magnetic field power source is switched off, it takes a finite interval of time for the curved space to return to flat space. In the mean time, spaceship is independent of curved space. It is therefore possible for the spaceship to proceed ahead receiving the action from the acceleration field. Namely, instantaneously switching off the magnetic field breaks the equilibrium state. Being independent of curved space, spaceship is subjected to the action of field during the finite interval to proceed ahead. In general, a body cannot move carrying, or together with, a field that is generated by its body. In other words, the body cannot move unless the body is independent of the field. In a surrounding region of spaceship, a magnetic field as an engine produces a curved space. By switching off the magnetic field, in an instantaneous transition interval which the curved space disappears and returns to flat space, the spaceship is independent of curved space. No interaction is present between curved space and spaceship.

Here, the switching on-off the magnetic field implies the following consideration. There exists the seed magnetic field to be compressed in the engine system. Using the magnetic flux - compression technology, we compress the seed magnetic field and produce the spatial curvature induced by compressed strong magnetic field (equivalent of switching on the magnetic field). The power source of spaceship is consumed in the work of compressing the seed magnetic field. After that, we switch off the magnetic field and this implies the shutting off the power source of the compressing magnetic field. Note that it does not simply switch on and off the current. This is because the magnetic field cannot be cut off by backlash current at the time of cutting current. So, the magnetic flux - compression technology becomes effective.

Figure 4 shows the propulsion principle of space drive propulsion. Now referring to Figure 4, we describe the propulsion mechanism in detail.

As previously described, the space has a finite strain propagation velocity, i.e., strain rate (= velocity of light "c"). Even if the magnetic field is switched off, the curved space reverting to flat space needs a finite time, that is, a length of curved space region divided by strain propagation velocity. The spaceship which exists in the curved space can be propelled by the thrust from the field of acceleration, when the curved space returns to flat space. The spaceship cannot be propelled while the magnetic field is switched on (Figure 4 (b)). This is because the spaceship produces the field of acceleration by itself and a state of equilibrium is held, that is, the action of magnetic field to space and reaction from space hold equilibrium state. However, when the magnetic field is switched off, the state of equilibrium is broken, and therefore, the spaceship can be propelled by the thrust from the field of acceleration (Figure 4 (c)), because the spaceship is independent of the field. Accordingly, this propulsion system is essentially defined as a pulse propulsion system. To propel the spaceship, the strained surface of space as shown in Figure 4 (a) is preferable in principle. Namely, a space with an anti-symmetric curvature is preferable, that is, flat space in A' region and curved space region in A region.

Practically, the spaceship can be propelled even by the strained surface of space having symmetric curvature, as shown in Figure 4 (b), (c). While the spaceship moves to A' region, the curved space in A' region has already returned to flat space as shown in Figure 4 (c). Therefore, the thrust from the acceleration field of A' region does not act on spaceship. In addition, this thrust is proportional to the mass of a volume element that exists in acceleration field. Although the mass of spaceship exists in the A region, the mass of spaceship does not exist in the A' region, therefore the thrust from the A' region can be disregarded from the outset. It is consequently clear that the A' region does not counteract the thrust which results from the A region. The spaceship can get continuous thrust by repeating the pulse-like ON/OFF change of magnetic field at high frequency.

The propagation velocity of the change from flat space to curved space and the propagation velocity of the change from curved space to flat space are both the same, i.e., the velocity of light "c". This is true for both the A' region and A region. Furthermore, the time interval in which the curved space returns to flat space is the same for both the region A' and A. After being accelerated in the A region, the spaceship proceeds into the A' region. Meanwhile, since the curved space in the A' region returns to flat space, the acceleration in the A' region becomes zero. Further, the spaceship has its mass "m" mainly in the A region in the meantime and is subjected to the thrust given by $f_{(A)} = m\alpha$.

Conversely, the spaceship has not yet its mass appreciably in the A' region, and is subjected to the thrust given by $f_{(A')} = 0$. Therefore, the reverse thrust in the A' region does not exist from the outset. After all, we do not need the anti-symmetric curvature as shown in Figure 4 (a).

Space Drive Propulsion

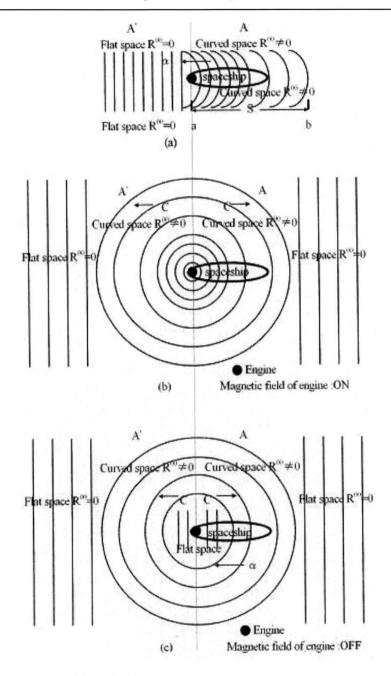

Figure 4. Propulsion principle.

It should be noted that the reverse thrust does not occur when the magnetic field is switched off. Someone is bound to think that a switching on and off of an intense magnetic field will make some sort of oscillatory spatial curvature and any net forward motion cannot be imparted to spaceship by this oscillating curvature. The magnetic field is switched on with a sufficient time in order to produce the curved space region shown by Figure 4 (a). In contrast to this, the magnetic field switching off time (t_{OFF}) is much shorter than the magnetic field switching on time (t_{ON}), i.e., $t_{OFF} \ll t_{ON}$. Therefore, the curved space region in opposite direction is small enough to be ignored and it cannot produce a reverse acceleration. Because the intensity of acceleration produced in curved space is proportional to both spatial curvature and the size of curved space. Since the region of reverse curvature is very small, the reverse thrust does not occur when the magnetic field is switched off. In addition, we can say that the amount of deviation of curved space in opposite direction is smaller than that of curved space in normal direction. Therefore, the spaceship can be propelled in a single direction.

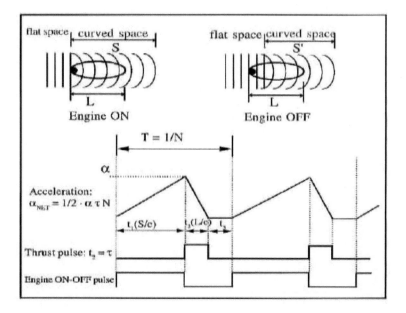

Figure 5. Timing chart of thrust pulse.

Figure 5 shows the timing chart of thrust pulse. The net acceleration α_{NET} is given by

$$\alpha_{NET} = 1/2 \cdot \alpha \tau N, \quad t_1 = S/c, \quad t_2 = \tau = L/c, \quad N = 1/(t_1 + t_2 + t_3) \, , (55)$$

where α = acceleration obtained by Eq. (51) or Eq. (54), S = length of curved space region (m), L = length of spaceship (m), c = speed of light (i.e., strain rate of space) (m/s), τ = effective time of thrust (i.e., thrust pulse width) (s), N = pulse repetition frequency (Hz).

So, the acceleration can be also controlled by N.

4.5. SOME EVALUATIONS OF SPACE DRIVE PROPULSION

Here let us evaluate other features of space drive propulsion such as momentum conservation law, energy conservation law, and the feature of flight performance.

Momentum and Energy Conservation Law

The question is that if the spaceship moves forward, then what moves back? As previously mentioned, the mechanism of propulsion can be classified into two kinds, i.e., momentum thrust (reaction thrust) and pressure thrust. The momentum thrust based on momentum conservation law is widely used in the present propulsion systems. On the other hand, the propulsion mechanism of pressure thrust is explained as follows: the propulsion obtained by pushing or kicking a huge massive body such as wall and ground.

In this case, the wall or ground pushes it back conversely as an external force, i.e., reaction. For example, a man can move forward by pushing his sole to the ground. At the local system between man and ground, the ground is fixed and does not move. However, at the global system between man and the Earth, since the Earth kicked by his sole moves back very slightly, the momentum conversation law is satisfied. All the same, the velocity of the Earth is nearly zero, then we can say that the Earth is fixed. Since man cannot throw out the Earth, it is not appropriate to apply the momentum thrust. It is impossible in principle for the rocket to throw out a heavier mass than rocket itself.

Considering the above, let us now think of four-wheel drive motorcar as an example of pressure thrust. In the case of the accelerating four-wheel drive motor car, the wheel kicks (pushes) the ground by rotating, and the wheel is subject to friction force from the ground. These frictions become a propulsive force of the motor car, i.e., thrust. Namely, this is the propulsion mechanism on the four wheels that kick the ground. Since these frictions from the ground are external forces for the motor car, the momentum conservation law is not satisfied so long as there exists an external force. In addition, the exhaust gas from the motor car is disregarded as thrust. However, at the global system including the Earth, the momentum conservation law is satisfied but this does not make any sense.

If the ground continues to an infinite cosmic space, the motor car can always move on the ground. There is no significance in applying the momentum conservation law to the ground infinitely-spread as a global system. The propulsion mechanism of the motor car is not momentum thrust but pressure thrust.

Another example of the pressure thrust is that the pressure thrust due to the difference between the nozzle pressure of the rear engine and the pressure of the front part of the aircraft partly contributes as the thrust of the rocket and jet plane.

Now, concerning the space drive propulsion system, the propulsion mechanism is also a kind of pressure thrust. As mentioned previously, its propulsion principle is based on the fact that the space is an infinite continuum. We regard the present space as an elastic body described by solid mechanics rather than by fluid dynamics. It may be easy to understand that the spaceship moves by pushing space itself, that is, by being pushed from space. The expression of "moves by pushing space or being pushed from space" indicates that the spaceship produces a curved space region and moves forward by being subjected to the thrust from the acceleration field of curved space. As the motorcar moves by kicking the ground infinitely-spread, the spaceship moves by pushing the cosmic space infinitely-spread. The cosmic space as an infinite continuum may be deformed very slightly by being pushed, just like the Earth moves back very slightly by being kicked due to the motorcar. However, this pushing is absorbed by the deformation of space itself. The whole cosmic space is considered as being similar to the ground for kicking. Thus, since the space behaves like the elastic field, the stress between spaceship and space itself is the key of propulsion principle. Accordingly, the analogy of rocket which obeys the momentum conservation law in Newtonian mechanics is not adequate.

If the body (spaceship) in space region gets the energy and the momentum, it means that the outside of body (spaceship), i.e., space as a field just loses them. Such a continuity equation means the global physical quantity conservation law. And when the body (spaceship) interacts with the field (space), in order to conserve the energy and momentum as a whole, it is necessary for the field (space) itself to get the energy, momentum, and stress. That is, it is the fundamental concept of the field theory.

In general, the energy-momentum conservation law is described by the continuity equation of the flow of physical quantities between the internal region V surrounded by an arbitrary closed surface (i.e., spaceship) and its surrounding field (i.e., space), that is,

$$-\frac{\partial}{\partial t}\int_V u dV = \int_V \nabla S dV + \int_V fv dV, \qquad (56)$$

where u = energy density in the region (volume V), S = the energy flux of the field (the flow of energy per unit time across a unit area perpendicular to the flow), fv = the rate of doing work inside volume V.

Eq. (56) stands for the conservation law in the field.

The total energy as well as the total momentum remains unchanged. They merely stream from one part of the field to another, and become transformed from field-energy and field momentum into kinetic-energy and kinetic-momentum of matter and vice versa.

According to Relativity, these quantities are related by the continuity equation as follows:

$$\frac{\partial}{\partial t}T^{00} + \frac{\partial}{\partial x^i}T^{0i} = 0, \quad \frac{\partial}{\partial t}T^{i0} + \frac{\partial}{\partial x^j}T^{ij} = 0 \; ; namely, \quad T^{ij}_{\;\;;j} = 0, \; (57)$$

where T^{00} = energy density, T^{0i} = energy flux, T^{i0} = momentum density, T^{ij} = momentum flux.

We can apply Eq. (56) and Eq. (57) to space drive propulsion mechanism. The space drive is the propulsion system utilizing the properties of continuum of space, and the interaction between spaceship and outside of spaceship (i.e., surrounding field) is the basic concept. The energy flux T^{0i} carries the momentum density T^{i0} ($T^{0i} = T^{i0}$). There is an important theorem in mechanics, that is, whenever there is a flow of energy in any circumstance at all (field energy or any kind of energy), the energy flowing through a unit area per unit time, when multiplied by $1/c^2$, is equal to the momentum per unit volume in the space.

The engine system of spaceship operates, and in the process of generating spatial curvature by compressing magnetic flux, the compressing energy from power source flows out as a strain energy flux T^{0i} and is stored in the surrounding space as a strain energy density T^{00}. The strain energy flux T^{0i} is accompanied by the strain momentum flux T^{ij}, and the momentum flux is stored in the surrounding space as a strain momentum density T^{i0}.

Next, shutting off the engine system of spaceship and releasing the compression of magnetic flux, the strain energy density T^{00} and strain momentum density T^{i0} stored in the surrounding space flow into the area of spaceship and are transformed into kinetic-energy and kinetic-momentum of spaceship with loss during this process. The above-mentioned mechanism is an interpretation of space drive propulsion from the standpoint of energy-momentum conservation law.

Spaceship Flight Performance and Feature

The spaceship equipped with space drive propulsion system has the following features.

a) There is no action of inertial force because the thrust is a body force. Since the body force they produce acts uniformly on every atom inside the spaceship, accelerations of any magnitude can be produced with no strain on the crews.
b) The flight patterns such as quickly start from stationary state to all directions in the atmosphere, quickly stop, perpendicular turn, and zigzag turn are possible.
c) The final maximum velocity is close to the velocity of light.
d) Since the air around the spaceship is also accelerated with spaceship, the aerodynamic heating can be reduced even if the spaceship moves in the atmosphere at high speed (10-100km/s). However, it is expected that a plasma (ionized air) envelops the spaceship.
e) Due to the electromagnetic propulsion engine, there is no roar and no exhaust gas.
f) The engine and power source are installed in the spaceship, therefore it can fly in the atmosphere of a planet as well as in cosmic space.
g) By pulse control of magnetic field, the acceleration varies from 0G to an arbitrary high acceleration (e.g., 36G).
h) Deceleration is easy for re-entry into the atmosphere.

4.6. FINAL PHASE OF SPACE DRIVE PROPULSION THEORY: ACCELERATION INDUCED BY COSMOLOGICAL CONSTANT

In the latest cosmology, the terms vacuum energy and cosmological term "Λg^{ij}" are used synonymously. Λ is a constant known as the cosmological constant. The cosmological term is identical to the stress-energy associated with the vacuum energy. The properties of vacuum energy, i.e., cosmological term are crucial to expansion of the Universe, that is, to inflationary cosmology. The vacuum energy in de Sitter solution yields the result that the expansion accelerates with time and the total energy with a comoving volume that grows exponentially [13, 14, 15, 16, 17]. These facts are due to the elastic nature of the vacuum and support the basic concept of space drive propulsion system, that is, the space is an infinite continuum. According to the gauge theories, the physical vacuum has various ground states. The potential of vacuum has minima which correspond to the degenerate lowest energy states, either of which may be chosen as the vacuum. Whatever is the choice, however, the symmetry of the theory is spontaneously broken. The particular interest for cosmology is the theoretical expectation that at high temperatures, symmetries that are spontaneously broken today were restored.

The most general form of the gravitational field equations which include cosmological constant is given by

$$R^{ij} - \frac{1}{2} \cdot g^{ij} R = -\frac{8\pi G}{c^4} T^{ij} + \Lambda g^{ij}, \tag{58}$$

where R^{ij} is the Ricci tensor, R is the scalar curvature, G is the gravitational constant, c is the speed of light, T^{ij} is the energy momentum tensor, and Λ is the cosmological constant.

It is simple to see that a cosmological term Λg^{ij} is equivalent to an additional form of energy momentum tensor. The cosmological term is identical to the energy momentum tensor associated with the vacuum.

Here, if we multiply both sides of Eq. (58) by g_{ij}, we obtain

$$\frac{8\pi G}{c^4} T = R + 4\Lambda. \tag{59}$$

In empty space with all of the components of the energy momentum tensor are equal to zero, that is, $T^{ij} = 0$ and $T = 0$, from Eq. (59) and Eq. (58), we get the following respectively (see *Appendix: A < Example >*)

$$R = -4\Lambda, \ R^{ij} = -\Lambda g^{ij}. \tag{60}$$

The scalar curvature R (1/m²) is given by

$$R = R_i^{\ i} = g_{ij} R^{ij} = g_{00} R^{00} + g_{11} R^{11} + g_{22} R^{22} + g_{33} R^{33} \approx$$
$$\approx g_{00} R^{00} = -R^{00} \ (g_{00} \approx -1: weak \ field). \tag{61}$$

Hence, from Eq. (60), we get

$$R^{00} = 4\Lambda. \tag{62}$$

Eq. (62) means that the cosmological constant Λ generates the major component of curvature of space R^{00}. Therefore, the curvature of space is identical as the cosmological constant.

Now, concerning the de Sitter cosmological model with non-zero vacuum energy (i.e., cosmological constant), the de Sitter line element is written as

$$ds^2 = -(1-\frac{1}{3}\Lambda r^2)c^2 dt^2 + \frac{1}{1-\frac{1}{3}\Lambda r^2} dr^2 + r^2(d\theta^2 + \sin^2\theta d\varphi^2), \tag{63}$$

where the metrics are given by

$$g_{00} = -(1 - 1/3 \cdot \Lambda r^2), \ g_{11} = g_{22} = 1, \ g_{33} = 1/(1 - 1/3 \cdot \Lambda r^2),$$

$$other \ g_{ij} = 0. \tag{64}$$

The acceleration α of de Sitter solution can be obtained by combining Eq. (54) with Eq. (64)

$$\alpha = \frac{1}{3}c^2 \Lambda r \quad (1 > 1/3 \cdot \Lambda r^2) .\tag{65}$$

The acceleration induced by the cosmological constant is proportional to the distance "r" from the generative source, i.e., engine system. According to the gauge theories, the physical space as a vacuum is filled with a spin-zero scalar field, called a Higgs field. The vacuum energy fluctuates in proportion to the fluctuation of the Higgs field [15]. The vacuum potential (vacuum energy density: J/m^3) V (ϕ) is given by the vacuum expectation value ϕ of Higgs field, and we get the minimum of the vacuum potential V$_0$ (ϕ) as follows:

$$V_0(\phi) = \frac{\lambda}{4}\phi_0^{\,4} .\tag{66}$$

Here, λ is arbitrary Higgs self-coupling in the Higgs potential (λ is not known and is not determined by a gauge principle, presumably $\lambda \geq 1/10$).

Since the vacuum potential V$_0$ (ϕ) shall be invariant under the Lorentz transformation, the energy momentum tensor of vacuum T^{ij}$_{vac}$ is written in the form

$$T^{ij}{}_{vac} = V_0(\phi)g^{ij} .\tag{67}$$

The energy momentum tensor of vacuum exerts the same action as that for the cosmological term. It should be noted that T^{ij}$_{vac}$ is not energy momentum tensor for matter but the vacuum itself.

From Eq. (58) and above Eq. (67), as its metric source, 8πG/c^4 · T^{ij}$_{vac}$ = 8πG/c^4 · V$_0$ (ϕ)g^{ij} = Λg^{ij}, then we get

$$\Lambda = \frac{8\pi G}{c^4} V_0(\phi) = 2.1 \times 10^{-43} V_0(\phi). \tag{68}$$

In general, since the potential from its source is inversely proportional to the distance "r" from the potential source, assuming that the vacuum potential $V_0(\phi)$ in Eq. (66) is the energy source, the potential at distance "r" apart from its energy source is written in the form

$$V_0(\phi) \Rightarrow V_0(\phi)/r = \frac{\lambda}{4r} \phi_0^4. \tag{69}$$

Combining Eq. (68) with Eq. (69) yields:

$$\Lambda = 2\pi G \lambda \phi_0^4 / c^4 r. \tag{70}$$

Substituting Eq. (70) into Eq. (65), finally we get:

$$\alpha = \frac{2\pi G \lambda}{3c^2} \phi_0^4 = 1.6 \times 10^{-27} \lambda \phi_0^4. \tag{71}$$

Eq. (71) indicates that the vacuum expectation value ϕ_0 for the Higgs field (i.e., vacuum scalar field) produces the constant acceleration field. As a result, we found out that the acceleration becomes constant, that is, we can get rid of the tidal force inside of the spaceship. The scalar field ϕ can be thought of arising from a source in much the same way as the electromagnetic fields arise from charged particles. We have to search for the fields with the source. The size L of spaceship (i.e., length or diameter) is limited to the range r_S, where r_S is the range determined by the following: $V_0(r) \propto V_0/r_s \approx 0$ ($L = r_s$). Within the range of $L = r_s$, the tidal force in the spaceship and in the vicinity of spaceship can be removed, that is, the acceleration becomes constant within the range of a given region "r_s". The vacuum expectation value ϕ of Higgs field can be considered as the strength of the field, i.e., energy of the field.

Using Eq. (66), particular attention is paid to the role of ϕ_0. Here, only ϕ_0 is described in NATURAL UNIT ($c = \hbar = k_B = 1$). In general, natural units are used for the field of elementary particle physics or cosmology. Since the fundamental constants $\hbar = c = k_B = 1$ are used in this unit system, there is one fundamental dimension, energy, can normally be stated in GeV, that is, [Energy] = [Momentum] = [Mass] = [Temperature] = [Length]$^{-1}$ = [Time]$^{-1}$: in GeV.

GeV^4 implies energy density (J/m^3) in SI unit. GeV^3 implies number density ($1/m^3$).

The following relation: $1GeV^3 = 1.3 \times 10^{47} m^{-3}$ is used to convert from the natural unit system to SI unit system. The vacuum expectation value ϕ_0 of the present universe is said to be $\phi_0 \sim 10^{-12}$ GeV and $\phi_0^4 = 1 \times 10^{-46}$GeV^4, therefore substitution of Eq. (66) and Eq. (71) with setting $\lambda = 1$ gives: **V$_0$(ϕ) = 1/4·ϕ_0^4 = 0.5 × 10^{-9}J/m^3**, $\alpha = 1.6 \times 10^{-27} \phi_0^4 = 3.3 \times 10^{-36}$ m/s$^2 \approx 0$. Naturally, the acceleration induced by present cosmic space is zero. In addition, from Eq. (68) and Eq. (62), we get $\Lambda = 2.1 \times 10^{-43}$ V$_0$(ϕ) = 1.05 × 10^{-52}m^{-2}, R^{00} = 4.2 × 10^{-52}m$^{-2} \approx 0$. Therefore, the present cosmic space is flat space.

From Eq. (39), the value of R^{00} = 4.2 × 10^{-52}m^{-2} gives the magnetic field of B = 7.2 × 10^{-4}gauss (7.2 × 10^{-8}Tesla). This value of magnetic field agrees quite well with the value of the interstellar magnetic field, i.e., ~10^{-5}gauss. As a temporary calculation, the vacuum expectation value ϕ_0 of present universe is excited and becomes $\phi = 6 \times 10^{-3}$GeV = 6MeV (from ϕ_0 to $\phi = \phi_0 + d\phi = \phi_0 + 6$MeV), similarly, we get the following: V$_0$ (ϕ) = 1/4· ϕ_0^4 = 6.7 × 10^{27}J/m^3, $\alpha = 1.6 \times 10^{-27} \phi_0^4 = 43.13$ m/s^2 = 4.4G, $\Lambda = 2.1 \times 10^{-43}$, V$_0$($\phi$) = 1.4 × 10^{-15}m^{-2}, R^{00} = 5.6 × 10^{-15}m^{-2}. From Eq. (39), the value of R^{00} = 5.6 × 10^{-15}m^{-2} gives the equivalent magnetic field of B = 2.6 × 10^{11}Tesla.

Phase Transition of Space

The space is a kind of continuum which repeats expansion and contraction. We assume that space as a continuum has two kinds of phases, that is, the elastic solid phase (i.e., Crystalline elasticity) like spring and the visco-elastic liquid phase (i.e., Rubber elasticity: = Entropy elasticity) like rubber. The elastic solid phase corresponds to the present universe and the visco-elastic liquid phase corresponds to the early universe. Further, we speculate that the space may get the phase transition easily by some trigger, i.e., excitation of space, and that the elastic solid phase of space is rapidly transformed to the visco-elastic liquid phase of space and vice versa. The space as a vacuum preserves the properties of phase transition even now. In general, the phase transition is accompanied by a change of symmetry. The phase transition has occurred from an ordered phase to a disordered phase and vice versa.

Figure 6 shows the phase transition of space.

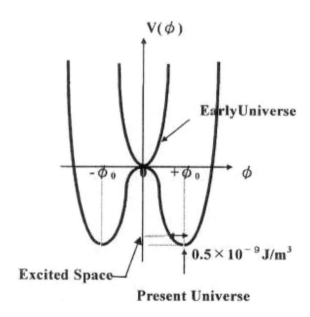

Figure 6. Phase transition of Space.

In a cosmological phase transition, the vacuum expectation value of scalar field ϕ is transferred from high-temperature, symmetric minimum $\phi = 0$, to the low-temperature, symmetry-breaking minimum $\phi = \pm\phi_0$. Accordingly, the phase transition is basically related to the spontaneous symmetry breaking, and it is considered that above-stated phenomenon is the fundamental property of space.

Now, referring to Figure 6, the vacuum expectation value of scalar field "$\pm\phi_0$" indicates the present true vacuum (present universe), and "$\phi = 0$" indicates the metastable false vacuum in early universe. Even if $\phi = \pm\phi_0$ had such a small vacuum potential value (0.5×10^{-9} J/m³), we would expect quantum fluctuations to push ϕ sufficiently far out on the potential from $\phi = \pm\phi_0$ to near the $\phi = 0$ by a trigger. Since the potential $V(\phi)$ (J/m³) means the energy density of vacuum corresponding to the expectation value of ϕ, the value of $V(\phi)$ directly contributes to the cosmological term. The change in ϕ gives the change in $V(\phi)$. As a result, the control of fluctuations of scalar field ϕ (i.e., coherent small oscillations of scalar field) affects the cosmological constant Λ. The enormous vacuum energy of the scalar field then exists in the form of spatially coherent oscillations within the field. As shown in Figure 6, a quantum fluctuations to push ϕ sufficiently by a trigger gives rise to a large perturbation of vacuum energy. Raising the vacuum potential may produce a large vacuum energy either through quantum or thermal tunneling, that is, pushing $+\phi_0$ by some trigger gives rise to a large perturbation of vacuum energy. Therefore, by taking above mechanism used as an unknown technology, we may produce a large cosmological constant in a local space, i.e., curvature. Here, the excitation of space means that the value of vacuum expectation value ϕ is pushed up slightly from its present value $\phi = +\phi_0$ and therefore the vacuum potential $V(\phi)$ is slightly raised.

Finally, we show the summary of space drive propulsion system in Figure 7.

SPACE DRIVE PROPULSION SYSTEM

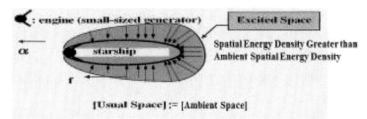

- Curvature of SPACE (R^{00}) plays a significant role for propulsion theory (Y.Minami:1988).

$$F^i = m\sqrt{-g_{00}}\,c^2\,\Gamma^i_{00} = ma^i = m\sqrt{-g_{00}}\,c^2\int_a^b R^{00}(x^i)dx^i$$

$$R^{00} = \frac{4\pi G}{\mu_0 c^4}\cdot B^2 \quad \text{Both strength of curvature and its extent (volume) are important.}$$

- Acceleration induced by de Sitter solution is found in 1996 by Minami : constant acceleration α (i.e. no tidal force inside of the starship).

$$\alpha = \frac{2\pi G\lambda}{3c^2}\phi_0^{\,4} = 1.6\times 10^{-27}\lambda\phi_0^{\,4}$$

Φ_0: non-zero vacuum expectation value of field

Figure 7. Summary of Space drive propulsion system.

4.7. RELATION BETWEEN ALCUBIERRE'S WARP DRIVE AND MINAMI'S SPACE DRIVE

As Gregory L. Matloff stated in his book "DEEP-SPACE PROBES" [18] as follows:
[…Somewhat more immediate are suggestions that we might create an artificial singularity using means other than gravity. Miguel Alcubierre and Yoshinari Minami have independently suggested that we might do this using magnetic field many orders of magnitude greater than those produced on the Earth – even greater than those at the surface of a neutron star or exotic fields

that might be manifested from the universal vacuum. Alcubierre's and Minami's ship (if possible) would be pushed or pulled through the Universe by a bubble of warped space-time...], these propulsion theories are well said to be alike.

Further, Edward J. Zampino (NASA Lewis Research Center) states their concept from the viewpoint of energy mainly in his paper entitled "Critical Problems for Interstellar Propulsion System" [19].

In conclusion, both propulsion theories are identical concept from the perspective that they are based on General Relativity and use the idea regarding distortion of space.

However, Alcubierre's warp drive is not manifest for its propulsion principle; there is no mechanism that how local distortion of space-time such as expansion space-time metric or contraction space-time metric create the thrust. Furthermore, Alcubierre's warp drive is a kind of non-used wormholes navigation theory for the purpose of interstellar travel [20].

While, Minami's space drive is manifest for its propulsion principle; there is obvious mechanism that the geometrical structure of space curvature creates actual force as thrust.

4.8. SPACE DRIVE PROPULSION FROM THE ASPECT OF COSMOLOGY

In previous sections 4.2−4.6, we ran over the propulsion theory of the space drive propulsion system. However, in this section, we explore the possibility that the expanding space generates thrust using the cosmology. That is, we make a study about the propulsion principle from the aspects of cosmology, especially considering the latest expanding universe theory of Friedmann, de Sitter, and inflationary cosmological model. Concerning equations using in this section, please refer to the textbook of Cosmology [13, 16, 17].

Expanding Space in Cosmology

The inflationary universe shows rapid expansion of space based on the phase transition of the vacuum exhibited by the Weinberg-Salam model of the electroweak interaction. The vacuum has the property of a phase transition, just like water may become ice and vice versa. Referring to Figure 8, shortly subsequent to the BigBang, early space is liquid like water, then during rapid expansion, space becomes solid like ice by decreasing temperature. This shows that a vacuum possesses a substantial physical structure such as the material. It coincides with the precondition of a space drive propulsion principle. In general, phase transitions are associated with a spontaneous loss of symmetry as the temperature of a system is lowered. For instance, the phase transition known as "freezing water", at a temperature T > 273K, water is liquid. Individual water molecules are randomly oriented, and the liquid water thus has rotational symmetry about any point; in other words, it is isotropic. However, when the temperature drops below T = 273K, the water undergoes a phase transition, from liquid to solid, and the rotational symmetry or molecular geometry of the water is lost. The water molecules are now locked into a 'solid' crystalline structure, and the ice no longer has rotational symmetry about an arbitrary point. In other words, the ice crystal is anisotropic, with preferred directions corresponding to the crystal's axes of symmetry [16].

Supposing that the universe expands, and then what form can the metric of space-time be assumed if the universe is spatially homogeneous and isotropic at all time, and what if distance is allowed to expand as a function of time? The metric they derived is called the Robertson-Walker metric. It is generally written in the form:

$$ds^2 = -c^2 dt^2 + a(t)^2 \left(\frac{dr^2}{1-Kr^2} + r^2(d\theta^2 + \sin^2\theta d\varphi^2) \right), \quad (72)$$

where $a(t)$ is the scale factor that describes how distance grows or decreases with time; it is normalized so that $a(t_0) = 1$ at the present moment. K is the curvature that takes one of three discrete constant values: $K = 1$ if the universe has positive spatial curvature, $K = 0$ if the universe is spatially flat, and $K = -1$ if the universe has negative spatial curvature.

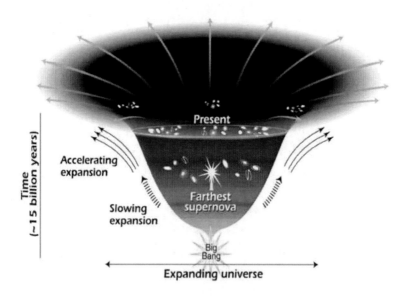

Figure 8. Expanding universe (adapted from Marc G. Millis).

The value of scale factor $a(t)$ is obtained by substituting the Robertson-Walker metric for the following gravitational field equation:

$$R^{ij} - \frac{1}{2} \cdot g^{ij} R = -\frac{8\pi G}{c^4} T^{ij} + \Lambda g^{ij}, \tag{73}$$

where R^{ij} is the Ricci tensor, R is the scalar curvature, G is the gravitational constant, c is the velocity of light, T^{ij} is the energy momentum tensor, and Λ is the cosmological constant.

That is, from the Robertson-Walker metric of Eq. (72), the Riemannian connection coefficient, the scalar curvature R, the Ricci tensor R^{ij} are obtained, and then substituting their value for Eq. (73), we get Eq. (74) as the case of $i = 0, j = 0$. Here ε is the energy density of space, $\dot{a}(t) = da(t)/dt$.

$$\frac{\dot{a}(t)^2}{a(t)^2} = \frac{8\pi G}{3c^2}\varepsilon - \frac{c^2 K}{a(t)^2} + \frac{1}{3}\Lambda c^2. \tag{74}$$

The Eq. (74) is called as the Friedmann equation and dominates the law of an expanding universe.

In a spatially flat universe ($K = 0$) and no cosmological constant ($\Lambda = 0$), the Friedmann equation takes a particularly simple form:

$$\frac{\dot{a}(t)^2}{a(t)^2} = \frac{8\pi G}{3c^2}\varepsilon. \tag{75}$$

From $\dfrac{\dot{a}(t)}{a(t)} = \sqrt{\dfrac{8\pi G}{3c^2}\varepsilon}$, $a(t)$ is obtained as the following:

$$a(t) = a_0 \exp\left[\left(\frac{8\pi G \varepsilon}{3c^2}\right)^{\frac{1}{2}} t\right] = a_0 \exp\sqrt{\frac{\Lambda}{3}}\, ct. \tag{76}$$

Here, from Eq. (68):

$$\Lambda = \frac{8\pi G}{c^4}V_0(\phi) = \frac{8\pi G}{c^4}\varepsilon. \tag{77}$$

So, we used the relation of $\varepsilon = \dfrac{c^4 \Lambda}{8\pi G}$ from Eq. (77).

A spatially flat universe with the energy density ε is exponentially expanding. Such a universe is called a *de Sitter universe*. Even if there is no cosmological constant Λ from the outset, in the nature of things, expanding universe is indicated by General Relativity. In initial assumptions, the energy density ε is considered as matter. At the present day, the energy density ε can be considered as the cosmological constant Λ.

Although the Friedmann equation is indeed important, it cannot, all by itself, indicate how the scale factor $a(t)$ evolves with time. We need another equation involving a and ε if we are to solve for a and ε as functions of time (t). As is well known in cosmology, they are the fluid equation:

$$\dot{\varepsilon} + 3\frac{\dot{a}}{a}(\varepsilon + P) = 0, \tag{78}$$

and the acceleration equation, using pressure P of the contents of the universe:

$$\frac{\ddot{a}}{a} = -\frac{4\pi G}{3c^2}(\varepsilon + 3P). \tag{79}$$

The acceleration equation can be derived from both the Friedmann equation and the fluid equation. The fluid equation Eq. (78) is derived from $\nabla T^i{}_j = 0$. Thus, we have a system of two independent equations in three unknowns – the functions $a(t)$, $\varepsilon(t)$, and $P(t)$. To solve for the scale factor $a(t)$, energy density $\varepsilon(t)$, and pressure $P(t)$ as a function of cosmic time, we need another equation, that is, the equation of state

$$P = \omega\varepsilon, \tag{80}$$

where ω is a dimensionless number, and is generally considered to take $\omega = 0$; the contribution of matter, $\omega = 1/3$; the contribution of radiation, and $\omega = -1$; the contribution of cosmological constant Λ.

The time-varying function $H(t)$ is generally known as the "Hubble parameter", while H_0, the value of $H(t)$ at the present day, is known as the "Hubble constant". Hubble parameter $H(t)$ is shown as

$$H(t) = \frac{\dot{a}}{a}. \tag{81}$$

So, the Friedmann equation evaluated at the present moment is

$$H_0^2 = \frac{8\pi G}{3c^2}\varepsilon_0 - \frac{c^2 K}{a_0^2}, \tag{82}$$

using the convention that a subscript "0" indicates the value of a time-varying quantity evaluated at the present (see Eq. (74)).

Incidentally, from $\nabla T^i{}_j = 0$, the following equation, i.e., the fluid equation, is obtained as time component (j = 0):

$$0 = \nabla_i T^i{}_0 = -\frac{d\varepsilon(t)}{dt} - 3\frac{\dot{a}(t)}{a(t)}\left(P(t)+\varepsilon(t)\right) = -\frac{1}{a(t)^3}\left[\frac{d(\varepsilon(t)a(t)^3)}{dt} + P(t)\frac{d(a(t)^3)}{dt}\right]. \tag{83}$$

The energy momentum tensor $T^i{}_j$ is defined as the following, assuming the fluid of space:

$$T^i{}_j = \begin{pmatrix} -\varepsilon(t) & 0 & 0 & 0 \\ 0 & P(t) & 0 & 0 \\ 0 & 0 & P(t) & 0 \\ 0 & 0 & 0 & P(t) \end{pmatrix}, \tag{84}$$

where energy density $\varepsilon(t)$ and pressure $P(t)$ are a function of cosmic time.

Now, regarding the cosmological constant Λ as a kind of energy momentum tensor of fluid, the energy density ε and the pressure P of vacuum space give the following from Eq. (78) or Eq. (83):

$$\varepsilon_\Lambda + P_\Lambda = 0. \tag{85}$$

Because, if the cosmological constant Λ remains constant with time, then so does its associated energy density ε_Λ, namely energy density ε_Λ is constant. The fluid equation Eq. (78) or Eq. (83) indicates that to have ε_Λ constant with time, the Λ term must have an associated pressure P_Λ.

Since $\dot\varepsilon = 0$ or $\dfrac{d\varepsilon(t)}{dt} = 0$, Eq. (85) is obtained. Thus, we can think of the cosmological constant as a component of the universe, which has a constant density ε_Λ and a constant pressure $P_\Lambda = -\varepsilon_\Lambda$.

Further, the energy density ε of the field of vacuum space is given by (see Eq. (77); dimension of vacuum potential V(ϕ) is J/m³)

$$\varepsilon_\Lambda = \frac{c^4 \Lambda}{8\pi G} . \tag{86}$$

Accordingly, the pressure P of the field of vacuum space becomes from Eq. (85):

$$P_\Lambda = -\varepsilon_\Lambda = -\frac{c^4 \Lambda}{8\pi G} . \tag{87}$$

In the case of $\Lambda > 0$, the pressure P_Λ of the vacuum field in Eq. (87) indicates the negative pressure, i.e., repulsive force.

Applying the value of $\Lambda = 2.1 \times 10^{-43} \times V_0(\phi) = 1.4 \times 10^{-15} m^{-2}$ (corresponding to: $\alpha = 1.6 \times 10^{-27}\phi_0^4 = 43.13$ m/s² $= 4.4G$) to Eq. (87), the pressure P of the field of vacuum space becomes 7×10^{27}Pa (7×10^{22} atm).

$$P_\Lambda = -\frac{c^4 \Lambda}{8\pi G} = \frac{(3 \times 10^8)^4 \times 1.4 \times 10^{-15}}{8 \times \pi \times 6.67 \times 10^{-11}} = \frac{81 \times 1.4 \times 10^{17}}{167.6 \times 10^{-11}} =$$

$$= 0.68 \times 10^{28} N/m^2 \approx 7 \times 10^{27} Pa$$

Applying the value of present universe of $\Lambda = 2.1 \times 10^{-43} \times V_0(\phi) = 1.05 \times 10^{-52} \text{m}^{-2}$ to Eq. (87),

$$P_\Lambda = -\frac{c^4 \Lambda}{8\pi G} = \frac{(3\times10^8)^4 \times 1.05 \times 10^{-52}}{8 \times \pi \times 6.67 \times 10^{-11}} = \frac{81 \times 1.05 \times 10^{-20}}{167.6 \times 10^{-11}} =$$

$$= 0.51 \times 10^{-9} \, N/m^2 \ .$$

The pressure P of the field of the vacuum space becomes 5×10^{-10}Pa $= 5 \times 10^{-16}$MPa $= 5 \times 10^{-15}$ atm ≈ 0.

Estimation of the Pressure of Vacuum Field

The pressure P_Λ of the vacuum field indicates the negative pressure, i.e. repulsive force.
Applying the value of $\Lambda=2.1 \times 10^{-43} V_0(\phi)=1.4 \times 10^{-15}\text{m}^{-2}$, the pressure of the field of vacuum space becomes 7×10^{27} Pa (7×10^{22} atm). [Excited Space]

$$P_\Lambda = -\frac{c^4\Lambda}{8\pi G} = \frac{(3\times10^8)^4 \times 1.4 \times 10^{-15}}{8 \times \pi \times 6.67 \times 10^{-11}} = \frac{81 \times 1.4 \times 10^{17}}{167.6 \times 10^{-11}} = 0.68 \times 10^{28} \, N/m^2 \approx 7 \times 10^{27} \, Pa$$

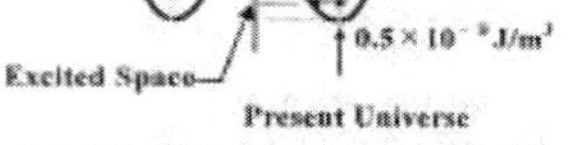

Applying the value of present universe of $\Lambda=2.1 \times 10^{-43} V_0(\phi)=1.05 \times 10^{-52}\text{m}^{-2}$, the pressure of the field of the vacuum space becomes 5×10^{-10}Pa$=5 \times 10^{-15}$ atm~0atm. [Present Space]

$$P_\Lambda = -\frac{c^4\Lambda}{8\pi G} = \frac{(3\times10^8)^4 \times 1.05 \times 10^{-52}}{8 \times \pi \times 6.67 \times 10^{-11}} = \frac{81 \times 1.05 \times 10^{-20}}{167.6 \times 10^{-11}} = 0.51 \times 10^{-9} \, N/m^2 \approx 5 \times 10^{-10} \, Pa \cong 0$$

Figure 9. Estimation of the pressure of vacuum field.

These are summarized in Figure 9.

Some early implementations of inflation associated the scalar field ϕ with the Higgs field, which mediates interactions between particles at energies higher than the GUT energy; however, to keep the discussion general, the field φ is now referred to as the inflation field. Generally speaking, a scalar field can have an associated potential energy $V_0(\phi)$.

Next we state about an inflationary cosmological model. In a cosmological context, inflation can most generally be defined as the hypothesis that there was a period, early in the history of universe, when the expansion was accelerating outward; that is, an epoch when $\ddot{a} > 0$.

The acceleration equation (79), $\dfrac{\ddot{a}}{a} = -\dfrac{4\pi G}{3c^2}(\varepsilon + 3P)$, tells us that $\ddot{a} > 0$ when $P < -\dfrac{\varepsilon}{3}$. Thus, inflation would have taken place if the universe were temporarily dominated by a component with equation of state parameter $\omega < -\dfrac{1}{3}$. Referring to Eq. (80), the usual implementation of inflation states that the universe was temporarily dominated by a positive cosmological constant Λ (with $\omega = -1$), that is, $P = -\varepsilon$.

Then Eq. (79) becomes

$$\frac{\ddot{a}}{a} = -\frac{4\pi G}{3c^2}(\varepsilon - 3\varepsilon) = \frac{8\pi G}{3c^2}\varepsilon. \qquad (88)$$

Substituting Eq. (86) into Eq. (88), thus had an acceleration equation that could be written in the form

$$\frac{\ddot{a}}{a} = \frac{8\pi G}{3c^2}\frac{c^4}{8\pi G}\Lambda = \frac{c^2}{3}\Lambda > 0. \qquad (89)$$

In an inflationary phase when the energy density was dominated by a cosmological constant, the Friedmann equation is described in

$$\frac{\dot{a}^2}{a^2} = \frac{8\pi G}{3c^2}\varepsilon - \frac{c^2 K}{a^2}. \qquad (90)$$

Setting flat space (K = 0), as well as Eq. (89),

$$\left(\frac{\dot{a}}{a}\right)^2 = \frac{\Lambda}{3}c^2. \qquad (91)$$

Since $\frac{\dot{a}}{a} = \sqrt{\frac{\Lambda}{3}}c$, we get

$$a = a_0 \exp\sqrt{\frac{\Lambda}{3}}ct. \qquad (92)$$

The scale factor grows exponentially with time. This result corresponds to Eq. (76). The vacuum space causes inflation by the energy of the vacuum and expands exponentially. The inflation mechanism brings up mini space to the macro space. Namely, the space has the property of exponential expanding by thermal energy [21, 22, 23, 24, 25].

Since the energy momentum tensor T^{ij} in the gravitational field equation (Eq. (73)) aims at matter, the gravitation arises between different matters. However, cosmological term "Λg^{ij}" in Eq. (73) implies that the force between the vacuum spaces, that is, repulsive force between one vacuum space and another vacuum space.

Space Propulsion Principle Brought about by Locally-Expanded Space

Next, we explain the space propulsion principle brought about by locally-expanded space in accordance with described above result. The vacuum space which envelops the spaceship is pushed by other expanding vacuum space, hence the spaceship is propelled by being pushed from the expanding vacuum space.

Concerning the propulsion principle for the space drive propulsion in the strict sense, it may be easy to understand that the spaceship moves by pushing space itself, that is, by being pushed from space. The expression of "moves by pushing space or being pushed from space" indicates that the spaceship produces a curved space region and moves forward by being subjected to the thrust from the acceleration field of the curved space.

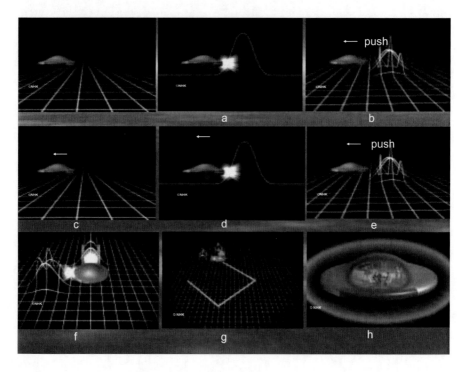

Figure 10. Explanation of spaceship operation (Motion of the spaceship using computer graphics).

Contrary to this, although it may be a loose expression, we can get an easy image of the propulsion principle: since the pressure of vacuum field in the rear vicinity of the spaceship is high due to an expansion of space, the spaceship is pushed from the vacuum field just like blowing up a balloon that can push an object.

Here, we explain the motion of the spaceship in detail using computer graphics as shown in Figure 10. For the sake of simplicity, the shape of the spaceship is an omni directional disk type.

As shown in Figure 10 (a), spaceship is able to permeate its local space with huge amount of energy in a certain direction; this energy should be injected at zero total momentum (in the spaceship-body frame) in order to excite the local space.

Then the excited local space expands instantaneously (Figure 10 (a), (b)). The space including the spaceship is pushed from the expanded space and advances forward (Figure 10 (b)). The space including the spaceship is propelled to the forward (Figure 10 (c)). Thus, this spaceship is accelerated to the quasi-speed of light by repeating the pulse-like on/off change of permeating its local space with huge amount of energy operation (Figure 10 (d), (e)). Changing a place to blow up, the spaceship can move with flight patterns such as quick start from stationary state to all directions, quickly stop, perpendicular turn, and zigzag turn (Figure 10 (f),(g)). There is no action of inertial force because the thrust is a body force. Since the body force they produce acts uniformly on every atom inside the spaceship, accelerations of any magnitude can be produced with no strain on the crews inside the spaceship. Namely, spaceship moves with the whole space around the spaceship, then, even if the spaceship flies about it very intensely, the spaceship holds the stopping state in moving space, and the crews are not shocked at all (Figure 10 (h)).

Furthermore, with the depiction of expanding space, referred to Figure 11, the spaceship is able to permeate its local space with a huge amount of energy in a certain direction; this energy should be injected at zero total momentum (in the spaceship-body frame) in order to excite local space. Then excited local space expands instantaneously. Space including the spaceship is pushed from the expanded space and advances forward. Since the pressure of the vacuum field in the rear vicinity of the spaceship is high due to an expansion of space, the spaceship is pushed from the vacuum field just like blowing up a balloon that can push an object (A→ B → C) as previously described.

We explored another possibility of a space drive propulsion principle where the locally rapid expanding space generates the thrust, using the cosmology, i.e., the latest expanding universe theory of Friedmann, de Sitter and inflationary cosmological model.

Thus, the space including the spaceship is pushed from the expanded space and advances forward. Although it may be a loose expression, we can get an easy image of creating a propulsion principle. The most important key seems to be the study of the structure of space that is derived from the expanding universe mechanics. In order to realize this result, we must discover the technology to excite and blow up space locally.

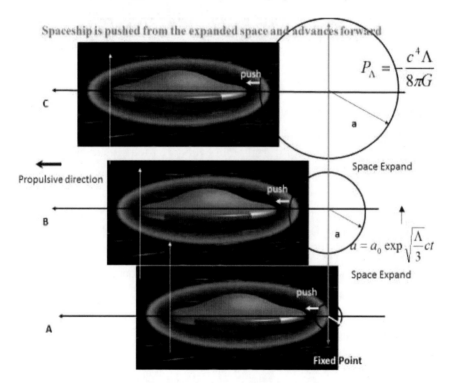

Figure 11. Propulsion principle based on expanding space.

References

[1] Minami, Y., "Space Strain Propulsion System", *16th International Symposium on Space Technology and Science (16th ISTS)*, Vol.1, 1988: 125-136.

[2] Forward, R.L. (Forward Unlimited, Malibu CA), *Letter to Minami, Y. (NEC Space Development Div., Yokohama JAPAN) about Minami's "Concept of Space Strain Propulsion System"*, (17 March 1988).

[3] Minami, Y., "Possibility of Space Drive Propulsion", *paper IAA-94-IAA.4.1.658, presented at 45th IAF Congress*, 1994.

[4] Hayasaka, H., "Parity Breaking of Gravity and Generation of Antigravity due to the de Rham Cohomology Effect on Object's Spinning", In *3rd International Conference on Problems of Space, Time, Gravitation*.1994.

[5] Huggett, S.A., and Todd, K.P., *An Introduction to Twistor Theory*, UK: Cambridge University Press, 1985.

[6] Pauli, W., *Theory of Relativity*, Dover Publications, Inc., New York, 1981.

[7] Minami, Y., "Space Drive Force Induced by a Controlled Cosmological Constant", *paper IAA-96-IAA.4.1.08, presented at 47th IAF Congress*, 1996.

[8] Minami, Y., "Spacefaring to The Farthest Shores - Theory and Technology of A Space Drive Propulsion System", *JBIS*, 50, 1997: 263-276.

[9] Minami, Y., "Conceptual Design of Space Drive Propulsion System", STAIF-98, edited by Mohamed S. El-Genk, *AIP Conference Proceedings 420, Part Three*, 1516-1526, Jan. 25-29, 1998, Albuquerque, NM, USA.

[10] Flügge, W., *Tensor Analysis and Continuum Mechanics*, Springer-Verlag Berlin Heidelberg New York, 1972.

[11] Fung, Y.C., *Classical and Computational Solid Mechanics*, World Scientific Publishing Co. Pre. Ltd., 2001.

[12] Minami, Y., "Continuum Mechanics of Space Seen from the Aspect of General Relativity — An Interpretation of the Gravity Mechanism", *Journal of Earth Science and Engineering* 5, 2015: 188-202.

[13] Kolb, E.W. and M.S. Turner., *The Early Universe*, Addison-Wesley Publishing Company, New York, 1993.

[14] Tolman, R.C., *Relativity Thermodynamics and Cosmology*, Dover Books, New York, 1987.

[15] Kane, G., *Modern Elementary Particle Physics*, Addison-Wesley Publishing Company, New York, 1993.
[16] Ryden, B., *Introduction to Cosmology*, Addison Wesley, 2003.
[17] Matsubara, T., *Introduction to Modern Cosmology Coevolution of Spacetime and Matter*, University of Tokyo Press, 2010.
[18] Matloff, G. L., *Deep Space Probes*, Springer, 2000; page 127 (Ch. 9: 9.4 'Cabbages and Kings': General Relativity and Spacetime Warps.
[19] Zampino, E.J., Critical Problems for Interstellar Propulsion Systems, Available from: ralph.open ─ aerospace.org/deep/repository/zampino2.pdf; website shown on Google, June 1998.
[20] Alcubierre, M., "The Warp Drive: Hyper-Fast Travel within General Relativity", Class. *Quantum Gravity* 11, L73-L77, 1994.
[21] Minami, Y., "Space drive propulsion principle from the aspect of cosmology", in: *STAIF (Space Technology & Applications International Forum)* II, Albuquerque, NM, Apr. 16-18, 2013.
[22] Minami, Y., "Space Drive Propulsion Principle from the Aspect of Cosmology", *Journal of Earth Science and Engineering* 3 (2013) 379-392. http://davidpublishing.org/
[23] Minami, Y., "Basic concepts of space drive propulsion—Another view (Cosmology) of propulsion principle—", *Journal of Space Exploration* (METHA PRESS), (2013) 106-115.
[24] Minami, Y., *A Journey to the Stars – By Means of Space Drive Propulsion and Time-Hole Navigation* ─ published in Sept. 1, 2014 (LAMBERT Academic Publishing; https://www.morebooks.de/store/gb/book/a-journey-to-the-stars/isbn/978-3-659-58236-3).
[25] Williams, C. (Editor); Minami, Y. (Chap.3); et al. *Advances in General Relativity Research,* Nova Science Publishers, 2015.

Chapter 5

ASTROPHYSICAL PROPULSION

Yoshinari Minami

5.1. ASTROPHYSICAL PHENOMENA

Overview

Here, astrophysical phenomena refer mainly accretion disk and astrophysical jet around black holes. Accretion disk is rotating gaseous disk with accretion flow, which form around gravitating object, such as white dwarfs, neutron stars, and black holes. At the present day, owing to the development of observational technology, it is believed that accretion disk causes the various active phenomena in the universe: star formation, high energy radiation, astrophysical jet, and so on.

It should be noted; these stars such as white dwarfs, neutron stars, and black holes have a strong magnetic field (10^8 Tesla-10^{11} Tesla). Matter falling onto an accretion disk around black hole is ejected in narrow jet moving at close to the speed of light like an accelerator.

Entity of the astrophysical jet is a jet of plasma gas from the active galactic nucleus (accretion disk in there). It is said that such astrophysical jet is held together by strong magnetic field tendrils, while the jet's light is created by particles revolving around these wisp thin magnetic field lines. Furthermore, since the system of black hole and accretion disk is like a gravitational power plants, the energy of the heat and the light are produced by the release of gravitational energy.

Figure 1 shows astrophysical jet and accretion disk around black hole.

Although the system of accretion disk and astrophysical jet around black holes are currently left many unresolved issues, the elucidation of these mechanisms and principles that are common to the entire universe may provide a new space propulsion principle. Especially, the breaking of magnetic field lines and magnetic field reconnection are possible to produce many kinds of charged particles such as electron-positron pairs. Generally, in a high-temperature plasma, electron-positron pairs are readily formed by collisions between the high energy protons, electrons, photons. Since the dynamics of the accretion disk has been decided by a magnetic field, it is important to solve the dynamics of the magnetic field.

The application of mechanism of accretion disk and astrophysical jet around black holes will lead to the concrete system design of propulsion engine and power source installed in space drive propulsion system [1, 2, 3].

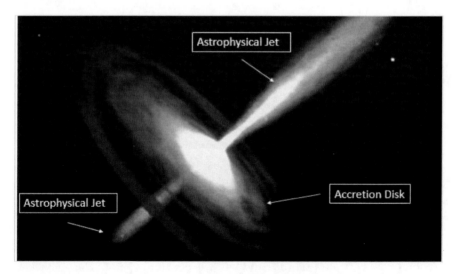

Figure 1. Astrophysical jet and accretion disk around black hole.
(https://pixabay.com/ja/%E3%83%96%E3%83%A9%E3%83%83).

Astrophysical Jet

The astrophysical jet is a narrow jetted plasma jet at high speed (100 km/s to near the speed of light) that emits in both directions vertically from accretion disk around the compact central object such as a neutron star or black hole. Its length is an enormous, long and narrow jet reaching from 1 light year—10 light years—1 million light years. A jet propagating at a speed close to the speed of light is called a relativistic jet (See Figure 1.).

The acceleration mechanism of the astrophysical jet and the collimation mechanism narrowing down to a long distance have been examined so far. They are due to thermal gas pressure, light radiation pressure, and magnetic field pressure. Currently, Radiative Acceleration model accelerated by the radiation field of the accretion disk and Magnetic Acceleration model accelerated by magnetic field penetrating the accretion disk are representative models. The high velocity, highly collimated gas streams - jets - raise two major problems, namely how the jet material is accelerated, and how it is collimated (Figure 2 (a)).

It is a collimation problem of how to narrow the jet narrowly, and the model of the jet acceleration mechanism is required to solve this collimation problem at the same time as well as acceleration. At the present time, the magnetic force model (magnetic centrifugal force and magnetic pressure) is regarded as the most dominant theory which solves the two problems of jet acceleration and collimation at the same time. That is, the accretion disk generates a helical magnetic field by twisting the magnetic field lines, accelerates by magnetic force, and narrows the jet by magnetic tension (pinch). The self-pinching force of magnetic field twisted by the rotation occurs naturally as a force to collimate the jet thinly (Figure 2 (b)) [4, 5, 6, 7, 8, 9, 10, 11, 12].

If the magnetic field lines are in the jet, there is a possibility that a strong magnetic field region is locally generated due to local turbulence and shock waves in the plasma.

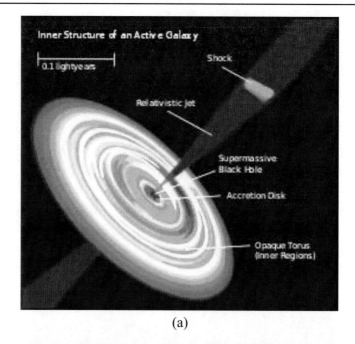

(a)

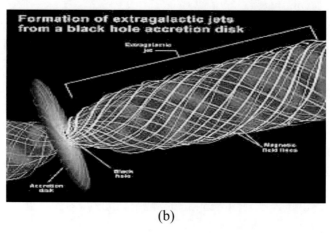

(b)

Figure 2. (a) Astrophysical Jet from an Accretion Disk of Black Hole; (b) Formation of Astrophysical Jet wound by twisting magnetic field lines.
(http://image.search.yahoo.co.jp/sear ch?rkf=2&ei=UTF-8&gdr=1&p= Astrophysical+Jet).

Acceleration and Collimation Mechanism of Astrophysical Jet

An astrophysical jet is a phenomenon often seen in astronomy, where streams of matter are emitted along the axis of rotation of a compact central object (such as a black hole or neutron star). Many stellar objects with accretion disks have jets. While it is still the subject of ongoing research to understand how jets are formed and powered, the two most often proposed origins are dynamic interactions within the accretion disk, or a process associated with the compact central object. When matter is emitted at speeds approaching the speed of light, these jets are called relativistic jets.

While it is not known exactly how accretion disks would accelerate jets or produce positron-electron plasma, they are generally thought to generate tangled magnetic fields that cause the jets to accelerate and collimate.

One of the astonishing properties of astrophysical jets is that they remain collimated over quite large distances. Again, magneto-hydrodynamic (MHD) processes seem to be most likely responsible for this behavior: the same pinch mechanism, which forced the plasma gas into a beam directed along the polar axis of the driving source, is also collimating the astrophysical jet further out. The idea of magnetic collimation of jets in the asymptotic regime (i.e., far from the driving sources) has been proposed first for galactic radio jets showed that any axisymmetric (nonrelativistic) magnetized wind will approach a cylindrically collimated structure, if the electric current carried by the flow is non-zero. The collimation mechanism is straightforward to understand for a current carrying flow: the current creates a magnetic field wrapping around the current via Ampère's law. The action of this (toroidal) field then pinches the current back to the flow axis via the Lorentz force. In the case of a vanishing current, the flow would still be paraboloidally collimated. The importance of magnetic fields for jet collimation is valid.

Figure 3 (a) shows General Relativistic MHD simulation on the interaction between ergosphere and magnetic field line of rotating black hole. A black hole dynamics has its plasma aligned to the interstellar magnetic field lines that thread through the equatorial plane of the accretion disk just outside the event horizon, so that the plasma evolves into accretion disk, which could be described as a condensate of electron-positron pairs. Since the plasma being highly conductive, it will be expected that the interstellar magnetic field lines will become frozen into that plasma which rotates within accretion disk, and as the accretion disk rotates it will drag and twist those magnetic field lines, pulling them together (See Figure 3(a) and Figure 4).

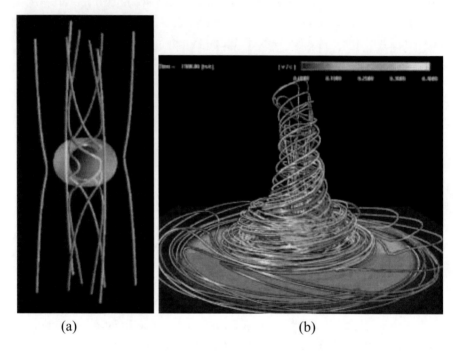

(a)　　　　　　　　　　　　(b)

Figure 3: (a) MHD jets from Kerr hole magnetosphere (Koide et al. 2002 Science); (b) KatoY, Mineshige, Shibata (2004); 3D sim. (ApJ). This toroidal field dominated jet is launched by magnetic pressure (similar to Shibata and Uchida 1985, Turner et al. 1999, Kudoh et al. 2002), and is also Similar to "magnetic tower" of Lynden-Bell (1996).

Figure 3 (b) shows a toroidal field dominated jet is launched by magnetic pressure. As the magnetic field penetrating the accretion disk is twisted, the energy of the magnetic field is accumulated, and at the same time that it propagates along the magnetic field lines, the jet ejects from the accretion disk, and not only the magnetic centrifugal force but also magnetic pressure contributes to acceleration of jet.

When the magnetic field is twisted in the direction of rotation by the actuation rotation of the plasma material, the twisted magnetic field acts like a spring to accelerate the plasma material further upward.

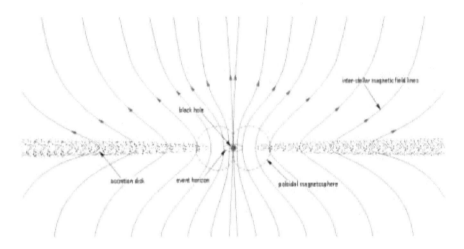

Figure 4. Interstellar magnetic field lines will become frozen into that plasma which rotates within accretion disk.

In other words, it is acceleration by magnetic pressure. If the jet is magnetically accelerated, the jet is expected to have a twisted helical magnetic field (Figure 2(b)). Furthermore, the twisted magnetic field acts like a rubber string, and the force of the rubber band shrinks (magnetic pinch) so that the flow of the plasma substance is directed in the rotation axis direction. This is the collimation of the jet by magnetic field. Collimation also occurs voluntarily in addition to acceleration in a model where a jet is driven from an accretion disk by magnetic field and rotation.

Even if the magnetic field penetrating the accretion disk is very weak, the rotation of the accretion disk causes the magnetic field to twist and increase more and more, and the energy is stored in the magnetic field to the same extent as the rotational energy of the accretion disk. Even in the case of a local magnetic field in the disk instead of the global magnetic field, the magnetic field is twisted in the disk, so that magnetic pressure is generated and it is possible to accelerate the jet [7, 8, 9, 10].

The energy of the magnetic field is increased by compressing the gas or stretching the magnetic field lines due to the plasma gas. The phenomenon in which energy stored in the form of a magnetic field is released locally and in large quantities in a short time is well known for solar flares. The acceleration mechanism for these jets may be similar to the magnetic reconnection processes observed in the Earth's magnetosphere and the solar wind. The energy of magnetic field is given by:

$$E_{mag} = \frac{B^2}{2\mu_0} \cdot \frac{4\pi}{3} R^3, \qquad (1)$$

where R is the radius of sunspot.

As magnetic rotation instability grows, magnetic field lines are stretched in the azimuth direction, the magnetic field is strengthened, and the magnetic energy exponentially increases.

As long as there is a weak magnetic field at the beginning, the magnetic field is amplified by magnetic rotation instability.

5.2. Background of Astrophysical Propulsion

Prologue

I was informed the book entitled "*Gravitational Manipulation of Domed Craft*; Potter, P. E. (2008) [13]" by Paul Murad (Retired Department of Defense).

After rough reading, I was surprised to see the shape of the engine and its arrangement method in the Figure 5. The engine shape is analogous to the engine shape in my patent: Space drive propulsion device, UK Patent GB 2262 844 B Patent published: 16.08.1995.

Figure 5 shows that there exist four engines or three engines bowling pin-like shape.

Astrophysical Propulsion 131

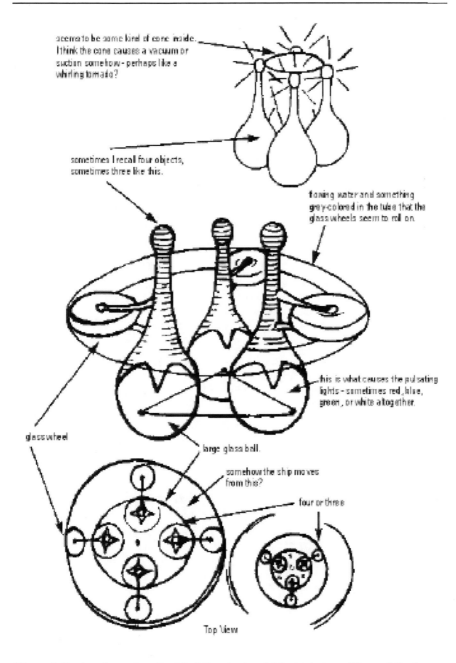

Figure 5. Engine shape described in "*Gravitational Manipulation of Domed Craft*; Potter, P. E. (2008)".

Next, I will briefly introduce the contents of the patent (Figure 6).

(12) **UK Patent** (19) GB (11) **2 262 844** (13) B

(54) Title of Invention

Space drive propulsion device

(51) INT CL⁵; H02N 11/00, B64G 1/40

(21)	Application No 9226804.4	(72)	Inventor(s) Yoshinari Minami
(22)	Date of filing 23.12.1992	(73)	Proprietor(s) NEC Corporation
(30)	Priority Data		(Incorporated in Japan)
	(31) 03356180		7-1, Shiba 5-chome
	(32) 24.12.1991		Minato-ku Tokyo
	(33) JP		Japan
(43)	Application published 30.06.1993	(74)	Agent and/or Address for Service
(45)	Patent published 16.08.1995		John Orchard & Co Staple Inn Buildings North High Holborn London WC1V 7PZ United Kingdom
(52)	Domestic classification (Edition N) H2A AKS3 AK121 AK125 AK220S AK808 B7G G43A		
(56)	Documents cited JP630073302 A		
(58)	Field of search As for published application 2262844 A viz: UK CL(Edition L) B7G, H2A AKRR AKR9 AKS3 AKS4 INT CL⁵ H02K 53/00 57/00, H02N 11/00 Online databases : WPI, CLAIMS, INSPEC updated as appropriate		

Figure 6. Patent application.

UK Patent GB 2262 844 B (Space Drive Propulsion Device)

Claims 1 and 2 that understand the whole function are described, and claims 3 to 6 are omitted here.

CLAIMS

1. A space drive propulsion device enclosing a hollow device region of a space and surrounded by a surrounding region of said space, said space being capable of having a field of curvature components, said device comprising:
    - magnetic field generating means for generating a controllable magnetic field which controllably generates said field of curvature components in said hollow device region and in said surrounding region; and
    - field control means for controlling said magnetic field generating means to make said magnetic field locally vary said curvature components substantially antisymmetric in said surrounding region.
2. A space drive propulsion device as claimed in Claim 1, said space drive propulsion device defining first through third axes of an orthogonal coordinate system, where in:
    - said magnetic field generating means comprises a plurality of magnetic field generating engines in predetermined relationships to said first through said third axes;

    each of said magnetic field generating engines comprising:
    - a spherical shell of a superconductive material enclosing a hollow shell space;
    - at least one controllable superconductor magnet in said hollow shell space to generate a pulsed magnetic field with a controllable pulse repetition frequency as at least a part of said controllable magnetic field;
    - said field control means individually controlling the superconductor magnets of said magnetic field generating engines to control the repetition frequency of the pulsed magnetic field generated by at least one of said superconductor magnets and to locally vary said curvature components substantially antisymmetric in said surrounding region.

Next, outline of engine shape and engine layout will be explained.

Figure 3 in patent shows top view of a spaceship. Six magnetic field generating engines 33(1), 33(2), 33(3), 33(4), 33(5), and 33(6) are mounted in a spaceship shell 31. Each engine 33 generates a strong magnetic field.

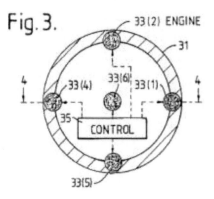

As described in Space Drive Propulsion (Chapter 4), the strong magnetic field generates spatial curvature and varies curvature components of a local field in a surrounding region around engine. Six engines are mounted in the front 33(2), back 33(5), left 33(4), right 33(1), up 33(6) and down 33(3) directions. This is because the spaceship moves in all directions.

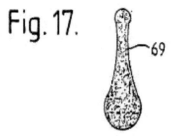

Figure 17 in patent shows an external view of functional engine 69 instead of engine 33. In Figure 17 (engine 69), the curvature of the space generated by the small sphere at the top is larger than the curvature of the space generated at the bottom large sphere. That is, the acceleration generated at the top is larger than that of the bottom. The shape of this engine

Astrophysical Propulsion

makes it possible to reduce the number of engines of 6 spheres (33) to 4 engines (69). Considering the vector synthesis of the thrust, the omnidirectional control of the spaceship can be executed with three engines (69) arrangement.

By using this engine (69), the number of engines of the spaceship will be four or three (same as Figure 5).

Figure 11 in patent shows the curvature of the space generated around the engine 33. In order to propel it in one direction, it is necessary to generate the curvature of an asymmetric space around the engine as shown in Figure 13 in patent.

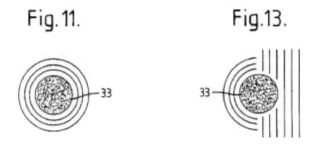

Figure 12 in patent schematically shows the coordination structure of the magnetic field generating the curvature of the asymmetric space of Figure 13 in patent.

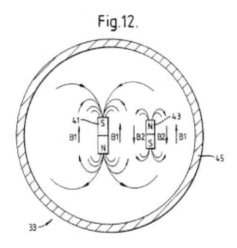

Figure 14 in patent shows a device to generate strong magnetic field using magnetic flux compression technology. This device uses a frozen-in magnetic field technology like a neutron star. A frozen-in magnetic field is a phenomenon that when an electrically conducting fluid such as liquid metal moves in the magnetic field, as if the magnetic flux or magnetic field adheres to the fluid, the magnetic flux is either deformed or intensified by the movement of fluid.

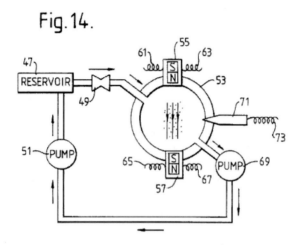

Fig.14.

A seed magnetic field to be compressed is added to the electrically conducting fluid in advance. After that, a small area of the fluid (liquid metal etc.) is irradiated with laser beam from all directions. The surface of fluid is heated by laser beam and ejects plasma. Subjected to the light pressure of laser beam and reaction by ejection of plasma, the fluid is very quickly concentrated. Namely, by cooperation of the light pressure and reaction of ejection of plasma, i.e., an ablation pressure are applied to the fluid, therefore an explosive magnetic flux compression is achieved. Consequently, since the fluid is contracted rapidly toward the center by light pressure and ablation pressure, the frozen-in magnetic flux of electrically conducting fluid is also contracted toward the center. Thus, the initial seed magnetic flux is compressed by laser implosion, and magnetic flux compression is achieved.

This operation resembles the state where the interstellar magnetic field lines is frozen in the plasma as shown in Figure 4.

Figure 15 in patent shows the seed magnetic field 59. Figure 16 in patent shows a small area of the fluid (liquid metal etc.) is irradiated with laser beam (75, 77) from all directions. Figure 16 in patent is similar to Figure 4.

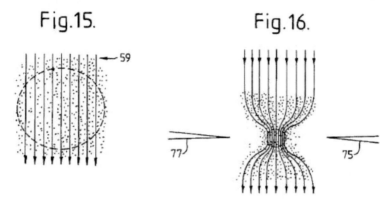

Book of *"Gravitational Manipulation of Domed Craft"*

According to the book entitled *"Gravitational Manipulation of Domed Craft*; Potter, P. E. (**2008**) [13]", the internal structure of the spacecraft is being introduced as shown in Figure 7, Figure 8, Figure 9 in detail. As well as UK Patent GB 2262 844 B, the strong magnetic field is also key item in this book's spacecraft.

However, in this book, the propulsion principle has not been explained clearly at all. Perhaps it seems that the propulsion principle that cancels the inertial force like gravity is not understood.

In the following section, we will nutshell the outline of amazing technology and scientific knowledge written in this book.

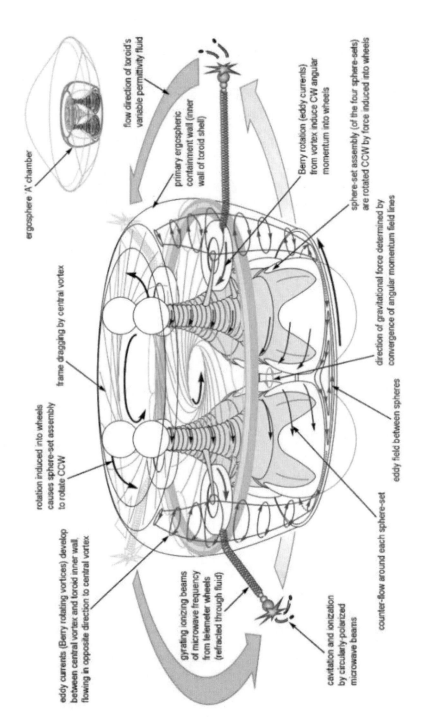

Figure 7. Spacecraft internal structure (1). (*Gravitational Manipulation of Domed Craft*; Potter, P. E. (2008)).

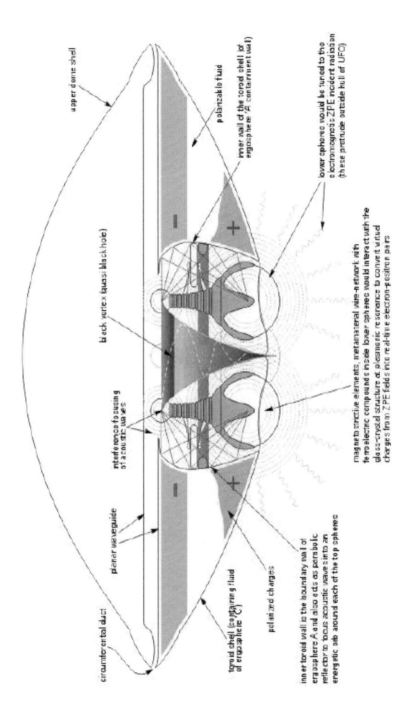

Figure 8. Spacecraft internal structure (2). (*Gravitational Manipulation of Domed Craft*; Potter, P. E. (2008)).

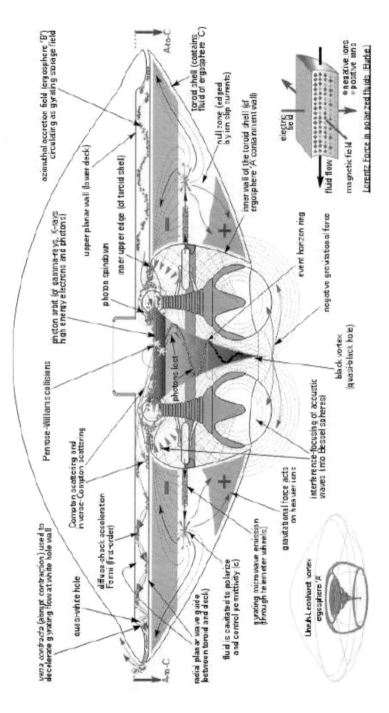

Figure 9. Spacecraft internal structure (3). (*Gravitational Manipulation of Domed Craft*; Potter, P. E. (2008)).

Figure 7, Figure 8 and Figure 9 show the internal structure of the engine and power source of the spacecraft described in this book.

Referring to these figures, the center black vortex (quasi-black hole) surrounded by the engine plays significant role for the strong magnetic field generating mechanism and generating power source, which utilizes the principle of the black hole and its accretion disk in the latest astrophysics.

Comparing the physical mechanisms shown in these figures with the astrophysics, it seems to be related to strong magnetic field, the ergosphere, the rotating vortex, the accretion disk's shearing field, the inertial frame dragging, the avalanche-ionization processes by magnetic flux line breaking and reconnection, the emission of synchrotron radiation, and the magnetic field anchor point, etc.

Especially, magnetic flux line breaking or magnetic field reconnection can be found working in all areas of space at all times.

Anyway, the astrophysical jet formation mechanism and the energy generation method by accretion disk centered on black hole hold the possibility of being applicable to a new propulsion system.

Now, since there are places where it is impossible to understand the functions and operations of each part shown in these figures for us, please refer to the original book (*Gravitational Manipulation of Domed Craft*; Potter, P. E. (2008);) for interested readers.

According to Figure 8 or Figure 9, spacecraft harvests energy from the ZPF field around the lower spheres.

The large lower spheres of the spacecraft are located so that they protrude outside of its metallic upper and lower domes. These lower spheres must therefore be positioned so as to have direct interaction between both the surrounding air (or incident electromagnetic radiation) and the internal electro-dynamic circuit of the spacecraft.

Since the charge density is inversely proportional to radius of curvature, the intensities of electric charge will be greater for the sphere of smaller curvature, that is, most of the electric charge collected or converted at the lower spheres will drift up to, and accumulate around the smaller top spheres (small spheres and lower spheres are interconnected).

Upper spheres could be continually replenished with abundant supplies of charged particles, from then being continually supplied by the lower spheres, and the ZPF energies converted through them from outside. The energy intensity exhibited by these upper and lower spheres around them must also be accompanied by a vast amount of charged particles. These charged particles would have to come in abundance through mechanism in operation close to the center of the spacecraft power drive system. These mechanism are induced by electron-positron generation from magnetic flux reconnection, electron-positron production through the virtual energy field in space, avalanche productions of more electron-positron pairs etc.

There will result two flows above and outside the toroidal shell, one of negative electrons circulating outward toward the outer rim, and the other flow will be the positive ions circulating inward over the inner edge of the toroid and toward the center of the spacecraft to where the sphere sets are. This mechanism involves in generating and mobilizing these electrical charges in and around the toroid fluid and even comparing these gyrating fields to an accretion disk which surrounds an astrophysical black hole. Negative charged particles will flow out from center (black vortex) while positive charged particles will flow center (black vortex).

Anyway, it is interesting for energy generating means as a power source and strong magnetic field generation.

Concerning the book, I do not know why the author (Paul E Potter) could have such new technical knowledge, but I feel that some descriptions adequately meet the latest astrophysical knowledge, furthermore, there is description about undiscovered knowledge and technology.

I think that some items seem to be misunderstood in part and not explained correctly, but the items pointed out are useful for future development. There is no explanation on the propulsion principle in particular, as to why thrust like gravity occurs and the spacecraft propels.

This propulsion principle is solved by applying the propulsion principle of space drive propulsion described in Chapter 4.

5.3. ASTROPHYSICAL SPACE DRIVE PROPULSION

In this section, we will introduce the promising space propulsion principle which is applied the latest astrophysical phenomena to a previously described space drive propulsion. This section is based on the matters of current astrophysics.

Energy Generating Method for Power Source

(a) Liquid Metal MHD Power Generation System Using Antiproton Annihilation Reactor

Any propulsion systems, i.e., not only conventional propulsion but also field propulsion (space drive propulsion), require huge energy sources due to their performance for producing high acceleration and high speed. This energy problem is common to all propulsion systems if high speed is required.

In general, a spaceship (mass of M) traveling at a speed V needs the kinetic energy of $E_K = \frac{1}{2} MV^2$. For instance, a spaceship traveling at a speed equal to 0.1 c has a specific kinetic energy equal to 450 TJ per kilogram (of spaceship mass). The required energy of spaceship of 100 ton at a speed of 0.1 c is 4.5×10^{19} Joules. Its power source in any propulsion system must provide huge energies, that is, $E = Pt$ (P is power in watts, t (s) is acceleration and deceleration time). Although this energy problem is common to all propulsion systems, the power source of field propulsion is based on allowable new advanced technology.

In the case of interstellar travel, the main problem could be summarized in one word: *Time*. Therefore, ultra high-speed (below quasi-light velocity) is required; as a consequence, huge energy is required. Especially, field propulsion system requires huge energy source due to their performance of high acceleration and high speed. The maximum speed of field propulsion is theoretically quasi-light velocity.

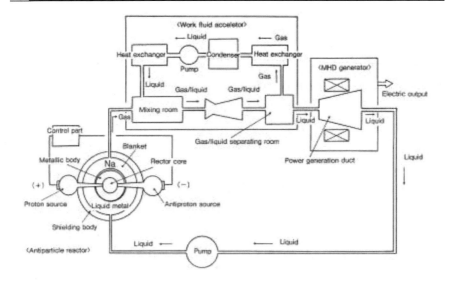

Figure 10. Antiproton annihilation reactor with liquid metal MHD generator.

How can we produce such a huge energy? Since we did not address the energy source for field propulsion in this book, we mention here about "New Energy Sources for Interstellar Travel or Intergalactic Exploration" in brief.

Minami proposed "Liquid Metal MHD Power Generation System Using Antiproton Annihilation Reactor" [14, 15]. Figure 10 shows the block diagram of Antiproton annihilation reactor with liquid metal MHD generator.

Two storage spheres which store protons and antiprotons face each other and beam tube from each storage sphere compose a globular reactor core at the center. The reactor core is surrounded by a metallic body with a high density and a blanket, and further by a shield body around it.

The blanket consists of liquid metal and its container. The metallic body and the blanket generate heat by absorbing γ quanta produced by the annihilation reaction. Generated heat is controlled by changing the flow of proton-antiproton source using the external electric field.

The liquid metal is heated in the antiparticle reactor, sent to mixing chamber in vapor phase at a high temperature and pressure, where the liquid metal is injected to make blow up from a nozzle as a high speed jet stream of gas-liquid two-phase flow. The fluid is separated into vapor and liquid in a gas-liquid splitter at the next stage, and only the liquid is made pass through the duct for MHD power generation to be employed for electric power generation.

The liquid metal from the MHD generator is sent again to the blanket of high temperature heat source (reactor) by means of a pump, and the above action is repeated.

Whereas for the vapor taken out at the gas-liquid splitter, it is deprived of heat through a heat exchanger, and returns to liquid metal at a condenser. The liquid metal is sent into the heat exchanger with a pump, and injected into the mixing chamber as liquid metal at a high temperature, repeating the above action. Liquid metal is best fit to extract heat stably from the reactor that is powerful heat source.

The detail is described in "Prometheus in Space: Survey Report of Research Committee on Functional New Material (JSUP, 1993); "4.4 Production of Antiprotons by Laser Accelerator"[14], and POSSIBILITY OF SPACE DRIVE PROPULSION (IAA-94-IAA.4.1.658) (JERUSALEM ISRAEL 1994) [15].

As a simple trial calculation, the $1m^3$ of hydrogen in the standard state contain 2.7×10^{25} hydrogen molecules, namely, 5.4×10^{25} protons.

Provided that the same number of protons and antiprotons exist in $1m^3$, the energy of $6.9 \times 10^{15} J/m^3 = 1.9 \times 10^9 kWh/m^3$ ($1kWh = 3.6 \times 10^6 J$) is obtained from the annihilation reaction. The heat source gained from an antiparticle annihilation reactor is supplied to an MHD (Magneto hydrodynamics) generator utilizing electromagnetic induction effect of an electromagnetic fluid.

Antiprotons are being produced today by using accelerators such as Bevatron not only in Japan but also in various countries in the word, but the amount of production is extremely little. The new methods of production of antiprotons and storage technology to avoid collision with residual gas and material are required.

Furthermore, the vacuum energy or so-called *dark energy* may be useful in the near future. Discovery and further investigation of new energy methods to power these propulsion systems is warranted.

Fortunately, there is a different energy generation mechanism in the universe than the one described above. That is, gravitational energy plays an important role.

(b) Energy Sources in the Universe

As is well known, the main energy source of stars is nuclear fusion, which takes place in their central regions. However, historically, chemical reactions or gravitational energy were considered to be the energy source. Although these energy conversion efficiencies are well known and are described in many documents, here, we compare chemical energy, nuclear energy, and gravitational energy from the viewpoint of energy-conversion efficiency using the rest-mass energy ($E = mc^2$) [6, 8].

Chemical Energy

For example, 1kg of coal is perfectly burned to produce 5000−8000 kcal of heat, while 1 kg of kerosene gives about 10000 kcal, which equals 4.2×10^7 J. The efficiency η_C of a chemical reaction is

$$\eta_C \approx 5 \times 10^{-10} (4.2 \times 10^7 / 1 \times (3 \times 10^8)^2). \tag{2}$$

This is too small to account for the solar and stellar energy sources.

Nuclear Energy

In hydrogen nuclear fusion, where four hydrogen atoms convert into one helium atom, the mass deficit per particle is $\sim 0.029/4 \doteqdot 7 \times 10^{-3}$. The efficiency η_N of hydrogen nuclear fusion is

$$\eta_N \approx 0.007. \tag{3}$$

This is sufficient for solar energy. If we consider normal stellar cases, nuclear energy is the best energy source. However, this is no longer the case for astrophysical objects containing compact objects, such as neutron stars or black holes.

Gravitational Energy

If an object of mass M gravitationally contracts from infinity to radius R, the gravitational energy released is approximately GM^2/R, while the material energy is Mc^2. The efficiency η_G of gravitational energy is

$$\eta_G \approx \frac{GM}{Rc^2} \approx 2 \times 10^{-6}. \tag{4}$$

For the Sun, this is also too small to be a solar energy source. In this way, the efficiency of gravitational energy is very small in the daily world.

However, in a compact object like a neutron star or a black hole, the efficiency of gravitational energy rapidly increases. In the case of a neutron star, if we apply solar mass to M and neutron star radius 10 km to R, the efficiency will be $\eta_G \sim 0.15$. Gas of accretion disk around neutron star can release gravitational energy of 15% of the mass of neutron stars.

In addition, efficiency is estimated to be $\eta_G \sim 0.42$ in the Kerr black hole that rotates on its axis.

Gravitational energy is the ultimate energy source from the viewpoint of efficiency. Moreover, what is shining with its gravitational energy is the accretion disk.

The mechanism by which the accretion disk shines is not nuclear fusion reaction like a star, but due to release of gravitational energy. The release of gravitational energy works only when black hole and plasma gas of accretion disk exist. When the plasma gas falls to the gravity well of the black hole, enormous energy can be extracted from the falling plasma gas.

When the rotating plasma gas of the accretion disk loses its angular momentum due to the viscosity of the gas and gradually moves to the inner trajectory, the gravitational energy becomes excessive by the difference of the gravity gradient of the black hole. Half of the surplus extra gravitational energy is spent to increase the rotation while the other half is used to heat the plasma gas of accretion disk through viscosity (friction). Finally, it is converted into light and released from the accretion disk (See Figure 11).

The viscosity in the accretion disk plays two important roles: transport of angular momentum and heating of the disk plasma. Here, we indicate the released gravitational energy.

The local potential energy dE released by accreting material dm falling in the potential well from r to $r-\Delta r$ is obtained as follows:

$$dE = E(r) - E(r - \Delta r) = \left(-\frac{GM}{r} + \frac{GM}{r - \Delta r} \right) dm = \frac{GMdm}{r^2} \Delta r, \qquad (5)$$

where M is the mass of black hole.

Half of this goes to rotation energy (kinetic energy) $E_{rotation}$, while the rest $E_{radiate}$ should be radiated away,

Figure 11. Mechanism of gravitational energy release from accretion disk. (http://blog.goo.ne.jp/mobarider/m/201507).

$$E_{rotation} = E_{radiate} = \frac{1}{2} \frac{GMdm}{r^2} \Delta r. \qquad (6)$$

When the mass of *dm* falls from the radius r to the radius $r-\Delta r$, half of the gravitational energy released by the orbit's drop is spent to increase the rotation, but the other half is used to heat the gas of accretion disk through viscous friction, finally it is converted to light and emitted from the surface of the accretion disk.

This is the mechanism of gravitational energy release: the gravity of the central compact object, the rotation of the surrounding plasma gas (angular momentum), the viscosity between the rotating adjacent gases dominate the accretion disk are key functions.

Here, we consider the flow of energy within the accretion disk. Gravitational energy once becomes thermal energy of ions and electrons, and finally released as light energy. The thermal energy of the ion moves to the thermal energy of the electron due to the Coulomb collision between the ion and the electron. The electrons that obtained the thermal energy collide with the ions again by Coulomb collision to emit energy photons (thermal bremsstrahlung), collide with magnetic lines of force to emit energy photons (synchrotron radiation), or collide with photons lose energy (inverse Compton scattering). Ions and electrons frequently collide with each other, thermal energy flows from heated ions to electrons, and the ions and electrons are in a state of thermal equilibrium at the same temperature. Furthermore, electrons are cooled by emitting photons.

Generally, positrons are generated in very high energy celestial phenomena. When an electron and a positron collide, it disappears and turns into energy. When electrons and positrons annihilate, a spectral line (electron-positron pair annihilation line) having a peak at 511 keV is generated. Such pair annihilation lines have been detected in many areas of the universe, from solar flares to interstellar spaces, neutron stars, black holes, active galactic nuclei.

In a high temperature plasma where the temperature reaches the threshold of 6 billion K, electron positron pairs ($e^-\ e^+$ pair) are easily formed by high energy protons and collisions between electrons and photons. It is thought that there is a high temperature plasma around the black hole where electron-positron pairs are generated.

Whereby on the inside of its ergosphere immediately surrounding the black hole, the freshly created UV and gamma-ray hard photons (by synchrotron radiation, inverse Compton scattering and pair annihilation) just outside the horizon, are drawn from the inside of that ergosphere down into the black hole and this spinning down results in being lost forever.

There will then occur ionization processes through which there will subsequently be generated even more electron-positron pairs. This transfer of energy from the black hole is what happens in descriptions of the accretion disk, and this stems from the fact that the flux lines of the co-mutual magnetic field will thread through both the ergosphere surrounding that black hole and the accretion disk.

Strong Magnetic Field Generation by Magnetic Field Line Break-Reconnection

In astrophysics, magnetic field reconnection can be working in all areas of space at all times. It works not only on the surface of the sun and in the sun's solar flares but also in rotating fields of accretion disks in space.

As is well known, magnetic field reconnection will provide copious productions of electron-positron charged particles, and will produce so much energy.

The magnetic reconnection is considered to be promising as a solar flare energy release mechanism, but it seems not necessarily clear.

The magnetic field lines play a role analogous to that of conducting wires in an electronic circuit. If the magnetic field lines are broken, then the entire voltage potential drop would be developed across that break.

Where the magnetic flux lines are snapped, the electric voltage that was being carried along the whole of that unbroken line would continue to traverse the gap between the snapped ends of that flux line, and would continue to increase in potential (by self-inductance).

The virtual particle of the vacuum will become unstable at that gap and will break down to transform those virtual particles into real electrically charged particles. When the breakdown in the vacuum occurs, avalanches of electron-positron pairs will be produced out of that of space. In space the potential drop threshold for when this effect takes place around the breaking of magnetic flux lines is said to be about 10^{12} volts. This particle producing mechanism can be found around Sun, where the chromosphere continually produces electrons and positrons by this method and then has to eject out from it around a million tons of those charged particles every second.

This magnetic field line reconnection process is working throughout the whole cosmos, and in active galactic nuclei which are some of all the astrophysical jets. The magnetic field reconnection system is regarded as one of the most efficient production method of charged particles in the galaxy. Around accretion disk the shearing-reconnection of strong magnetic field produces a dynamo effect which gives a rapid amplification of any incoming and smaller electrical field of charged particles (seed field). So, they will develop into much larger field and go on to accelerate particles which will collide with other particles, to produce more particles, and more collisions, which subsequently will lead to avalanche productions of more electron-positron pairs.

In impulsive reconnection mechanism which also occurs in accretion disk, these magnetic field line breaks and reconnections would lead to intense heating, and then anomalous dissipation of additional charged particles in localized regions around those flux lines. With the subsequent production of electron-positron pairs and then these particles feed back into the accretion disk to complete the dynamo effect. As the instabilities accumulate in energy and when the electron velocity exceeds that of ion-acoustic wave, a flurry of fast reconnection occurs resulting in explosive outburst of charged particles.

Anyway, electron-positron pair production comes from the breaking and reconnection of magnetic field lines. The breaking and remaking of magnetic field lines produces and then amplifies amounts of electrons and positrons from what some have called the 'empty' vacuum of space.

After all, the key is energy generation by magnetic field breaking and magnetic reconnection. Because mass production of charged particles by electron avalanche phenomenon and generation of electron positron pairs accompanying this can be utilized. The generation of a large amount of charged particles bring about the generation of a large current, and it is possible to generate a strong magnetic field from this large current.

A strong magnetic field is indispensable for energy generation and spatial curvature generation as propulsion drive.

Finally, concerning space drive propulsion, we explained Schwarzschild solution by strong magnetic field (as the first generation), de Sitter solution by excitation of space (as the second generation), and cosmological aspect (as the third generation), in chapter 4.

And, in this Chapter 5, we explained from the perspective of Astrophysical Phenomena (as the fourth generation).

Astrophysical Space Drive Propulsion is promising for propulsion engine and its power source. This is because the strong magnetic field and the power source for spatial curvature generation of the space drive propulsion system can be simultaneously solved by a single technology. The author is currently examining concrete system design and plan to publish it by article in the near future.

REFERENCES

[1] Minami, Y., "Spacefaring to The Farthest Shores - Theory and Technology of A Space Drive Propulsion System", *JBIS*, 50, 1997: 263-276.

[2] Minami, Y., "Space propulsion physics toward galaxy exploration", 2015, *J Aeronaut Aerospace Eng* 4: 2.

[3] Minami, Y., "Space Drive Propulsion Principle from the Aspect of Cosmology", *Journal of Earth Science and Engineering* 3, 2013: 379-92.

[4] Contopoulos, I., Gabuzda, D., Kylafis, N., Editors, *The Formation and Disruption of Black Hole Jets*, Springer, 2015.

[5] Dermer, C. D. and Menon, G., *High Energy Radiation from Black Holes*, Princeton University Press, 2009.

[6] Kato, S., Fukue, J. and Mineshige, S., *Black-Hole Accretion Disks ─ Towards a New Paradigm ─*, Kyoto University Press, 2008.

[7] Shibata, K., Fukue, J., Matsumoto, R., Mineshige, S., Editors, *Active Universe—Physics of Activity in Astrophysical Objects—*, SHOKABO, Tokyo, 1999.

[8] Fukue, J., *Shining Black-Hole Accretion Disks*, Pleiades PUBLISHING Co., Ltd., 2007.

[9] Mineshige, S., *Black Hole Astrophysics*, Nippon Hyoron sha co., Ltd., 2016.

[10] Koyama, K. and Mineshige, S., *Black Hole and High-Energy Phenomena*, Nippon Hyoron sha co., Ltd., 2007.

[11] Minami, Y., "A Journey to the Stars: Space Propulsion Brought About by Astrophysical Phenomena Such as Accretion Disk and Astrophysical Jet", Global Journal of Technology & Optimization, 2016, 7:2 DOI: 10.4172/2229-8711.1000197.

[12] Minami, Y., "Another Collimation Mechanism of Astrophysical Jet", *Journal of Earth Science and Engineering* 7, 2017: 74-89.

[13] Potter, P. E., *Gravitational Manipulation of Domed Craft*, Adventures Unlimited Press, 2008.

[14] JSUP, "Prometheus in Space: Survey Report of Research Committee on Functional New Material", 1993. (http://www.researchgate.net/publication/281423024).

[15] Minami, Y., "Possibility of Space Drive Propulsion", In *45th Congress of the International Astronautical Federation (IAF)*, (IAA-94-IAA.4.1.658), 1994. (http://www.researchgate.net/publication/280320680).

Chapter 6

GALAXY EXPLORATION: AN ATTEMPT TO BEGIN IT WITH AN INITIATIVE FOR INTERSTELLAR FLIGHT

Yoshinari Minami

STELLAR SYSTEM EXPLORATION

As for the number of short-distance stellar systems, 63 fixed stellar systems exist in less than 18 light years from the solar system, and 814 fixed stellar systems exist in less than 50 light years from the solar system. As an example of a short-distance stellar system, Alpha Centauri is the nearest star from the solar system in 4.3 light years, and the star Sirius is near the seventh nearest star in 8.7 light years. On the other hand, as an example of a long distance stellar system, the Pleiades star cluster exists 410 light years distance and Cygnus exists 1800 light years distance. Figure 1 shows a view of stellar system and Figure 2 shows Solar system and Interstellar system.

It is only the moon that mankind could reach. Apart from unmanned spacecraft, mankind only possesses like a boat that goes near the beach, and we do not yet have a ship with the ability to traverse the vastness of the universe.

Figure 1. Stellar system.

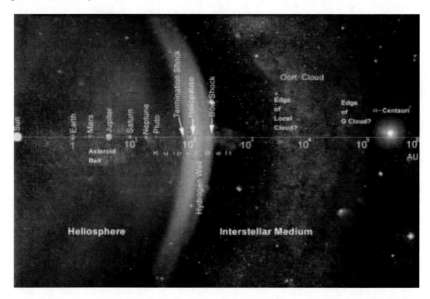

Figure 2. Solar system and Interstellar system (Under-chart source: R. Mewaldt & P. Liewer, JPL).

In the near future, the next target of the human beings who completed solar system exploration will go to stellar system exploration. Moreover, although the SETI project planned in each country until now is premised on actual existence of extraterrestrial intelligent life, the contact with them is made impossible by lack of the navigation theory and its technology which can conquer a huge distance between stars and Earth. That is, even if the speed of light is obtained, we have to spend the very long hours underway which will be required for tens of several year to hundreds years.

Considerable years are required even if it carries out an interstellar travel with the speed of light.

Recently, super-Earth as an extrasolar planet is discovered (Figure 3). The term *super-Earth* refers only to the mass of the planet, and does not imply anything about the surface conditions or habitability. However, the discovery of two new super-Earths around Gliese 581, both on the edge of the habitable zone around the star where liquid water may be possible on the surface are announced in April 2007. With Gliese 581c having a mass of at least 5 Earth masses, it is on the "warm" edge of the habitable zone around Gliese 581 (20 light years from the Sun) with an estimated mean temperature of −3 degrees Celsius. Its sister planet, Gliese 581d, does in fact lie within the star's habitable zone, with a mass of 7.7 Earths. Further, Tau Ceti at a distance of 12 light years from the Sun is said to be a terrestrial planet.

Figure 3. Super-Earth (Planetary orbits in the Gliese 581 system compared to those of our own Solar System) [Adapted from Wikipedia Gliese 581].

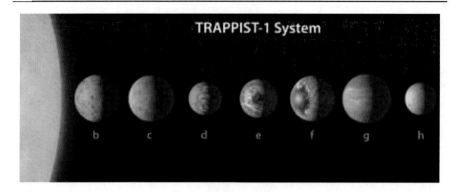

Figure 4. TRAPPIST-1 System (NASA Homepage).

Furthermore, just recently (Feb. 23, 2017), NASA announced that seven planets resembling the Earth were discovered at 39 light years distant from us. Three of these planets are firmly located in the habitable zone, the area around the parent star where a rocky planet is most likely to have liquid water.

Seven Earth-sized planets have been observed by NASA's Spitzer Space Telescope around a tiny, nearby, ultra-cool dwarf star called TRAPPIST-1 (Figure 4). Three of these planets are firmly in the habitable zone.

This exoplanet system at 39 light-years from Earth seems to be relatively close to us, but we cannot go to there due to lack of advanced propulsion system and navigation system. We need the practical space travel means which combines both space propulsion theory with space navigation theory.

G. Vulpetti discusses the problems and perspectives of interstellar exploration and shows how and why current physics does not allow real interstellar flight beyond the nearby stars, unless giant world ships are built and the concept of flight through generations is developed: two really formidable tasks indeed [1].

But on the other side, G. Vulpetti's "Conscious Life Expansion Principle (CLEP)" would entail, if true, the possibility of interstellar flight according to the fundamental laws of Nature.

Further, S. Santoli discusses the possibilities of space exploration that are envisaged on the basis of the novel emerging technologies that would lead to really autonomous robots [2].

Further, concerning interstellar travel, the method using a wormhole is well known; relying on space warps, such as for instance Wheeler-Planck Wormholes, Kerr metric, Schwarzschild metric, Morris-Thorne Field-Supported Wormhole based on the solutions of equations of General Relativity [3]. However, since the size of wormhole is smaller than the atom, i.e., ~10^{-35}m and moreover the size is predicted to fluctuate theoretically due to instabilities, space flight through the wormhole is difficult technically and it is unknown where to go and how to return. Additionally, since the solution of wormhole includes a singularity, this navigation method theoretically includes fundamental problems.

An interstellar travel within a human lifetime is impossible as long as we rely on only propulsion system. Even if the spaceship was to travel at the speed of light, an extremely long time are required. In order to conquer this huge distance and time, it is said to be that superluminal velocity becomes indispensable. Then, although the superluminal velocity tends to be expected simplistically, it becomes unreal expectation regrettably from the basic theory on physics and restrictions of the propulsion theory.

It is because there exists the wall of the velocity of light by the special relativity, any propulsion principle cannot exceed the velocity of light, and there is no energy source as the power source accelerated to that the wall of the velocity of light.

The special relativity theory acts correctly in the actual space and any propulsion theory cannot exceed the wall of the velocity of light.

Not propulsion theory but a new navigation theory becomes indispensable for the stellar system exploration as which the cruising range of a light-year unit is required. Although the navigation by the special relativity is well known as this kind of a navigation theory, it is the unreal navigation which does not become useful for the extreme time gap of global time and spaceship time as is well known as an Urashima effect (twin or time paradox). Even if we could reach to the fixed star in several years, what 100 years and what 1000 years had passed when it returned to the Earth of the hometown. It becomes a space travel of a one-way ticket literally.

Indeed, the problem of interstellar travel consists in much more in a navigation theory than in propulsion theory. The practical interstellar exploration combines both a propulsion theory with a navigation theory.

Therefore, also in the light of the Breakthrough Propulsion Physics Program by NASA, some reasonable theoretical speculations are necessary for trying to overcome the limits of the current physics.

H. D. Froning showed the rapid starship transit to a distant star (i.e., Instantaneous Travel) using the method of "jumping" over so-called time and space [4]. The detail is explained in Chapter 7 in this book.

Also, Y. Minami proposed Hyper-Space Navigation theory [5, 6]. The detail is explained in Chapter 8 in this book.

In the following chapters 7 and 8, promising approach of concepts regarding interstellar travel or intergalactic exploration is introduced.

REFERENCES

[1] Vulpetti G., "Problems and Perspectives in Interstellar Exploration", *JBIS*, 52, 1999: 307-323.

[2] Santoli S., Nano-to-Micro Integrated Single-Electron Bio-Macro Molecular Electronics for Miniaturised Robotic "Untethered Flying Observers", *Acta Astronautica*, 41, Nos. 4-10, 1997: 279-287.

[3] Forward, R.L., "Space Warps: A Review of One Form of Propulsionless Transport", *JBIS*, **42**, pp.533-542 (1989).

[4] Froning Jr, H.D., "Requirement for Rapid Transport to the Further Stars", *JBIS*, **36**, 1983: 227-230.

[5] Minami, Y., "Hyper-Space Navigation Hypothesis for Interstellar Exploration (IAA.4.1-93-712)", *44th Congress of the International Astronautical Federation (IAF)*,1993.

[6] Minami, Y., "Travelling to the Stars: Possibilities Given by a Spacetime Featuring Imaginary Time", *JBIS*, **56**, 2003: 205-211.

Chapter 7

RAPID TRANSIT BY FIELD PROPULSION TO DISTANT STARS

Herman D. Froning, Jr.

7.1. USE OF MINKOWSKI SPACES AND DE-BROGLIE WAVES TO EXPLORE: SOME FTL FLIGHT ISSUES IN HIGHER-DIMENSIONAL SPACETIMES

Some early faster-than-light investigations in the 1980's considered the idea that slower-than-light (STL) tardyons and faster-than-light (FTL) tachyons could body be solutions of Special Relativity, and the author was able to represent this idea geometrically by use of Minkowski spaces and de-Broglie waves. Here, the kinematics and dynamics of both tardyons and tachyons were described in a higher dimensional realm than that the lower-dimensional spacetime of tardyons. This section mentions some more work by the author which has used de-Broglie waves and Minkowski spaces to further explore some of the issues of FTL flight.

In the early 1980s speculative articles in physics journals introduced the idea of faster-than-light objects called "tachyons" and some authors [1] viewed tachyons as FTL solutions of Einstein's Special Relativity

(SR). For SR's Lorentz contraction factor $[(1-(V/c)^2]^{1/2}$ has real values only for speed V slower-than-light (STL). But, if multiplied by $i = [-1]^{1/2}$, this factor becomes $[(V/c)^2-1]^{1/2}$ with real values only for FTL speed. After some work, I was able to visualize both STL and FTL flight going on in a deeper x-ct-kτ realm than a shallower x-ct plane of existence of space and time. This deeper x-ct-kτ realm can be created by a vertical coordinate z = kτ, which is orthogonal to the x and ct coordinates which describe space travel (x) and time travel (ct) of STL tardyons on an x-ct space-time plane of existence. Figure 1 is an x-ct plane embedded in a higher-dimensional volume - an x-ct-ikτ realm that rises like a sky above x-ct space-time.

On the right in Figure 2 is a de-Broglie wave-packet approximation of a ship's vibrational state at slow STL speed, and on the left is its nearly-orthogonal packet state at fast FTL speed. Thus, an undulating, tardyon wave packet in x-y space-time is seen becoming a fast, faster-than-light tachyon that is undulating in a nearly-vertical orientation in a higher-D realm.

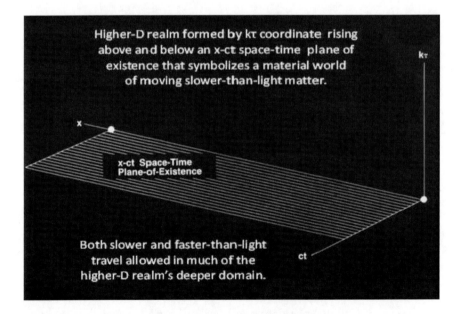

Figure 1. Lower-dimensional x-ct plane of existence of slower-than-light (STL) tardyons embedded in a deeper realm with room for both tardyons and swift faster-than-light (FTL) tachyons.

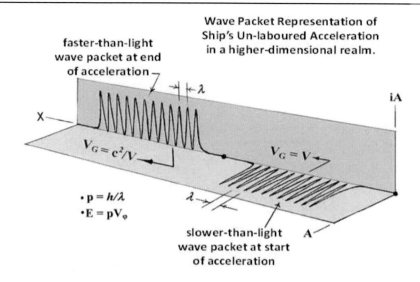

Figure 2. de-Broglie wave-packet approximation of a ship's vibrational state at slow STL speed, and on the left is its nearly-orthogonal packet state at fast FTL speed.

Stress-free flight requires ship momentum p to stay un-changed during acceleration. So a wave packet wavelength λ (which is proportional to **p**) must be un-changed during acceleration. The "phase velocity" V_φ of individual undulations of a tardyon wave-packet is FTL speed c^2/V_0 (where V_0 is STL). Initial wave packet group velocity V_G of the packet itself is V_0. In this example, packet V_G at maximum speed is c^2/V_φ whereas V_φ at the end of packet acceleration is the same as packet's initial V_G. The wave packet's vibration plane rotates in the higher x-ct-ikτ realm with change in packet speed V_G/c relative to Earth. Angle ψ of this plane relative to horizontal is $\tan^{-1}(V_G/c)$. So undulations are horizontal at zero V_G. But this vibratory plane rotates towards vertical with increasing V_G. So, vibrations are near-horizontal at slow initial ship speed and near-vertical at fast final ship speed.

My view of tardyons and tachyons as wave packets undulating in different planes in a higher-D realm than space-time was published [1] over objections of 2 reviewers who believed my tachyons were too "metaphysical". But my tachyons in a deeper realm avoided "causality" paradoxes that tachyons in a lower-D space-time suffered. For a tachyon in space-time could be seen by one observer as going forward-in-time, but to another at much different speed and distance it could be seen moving backward in time.

My wife Irina visited an office with a UFO book [2] on a table, and while idly thumbing through it, she saw the word "tachyon" (a strange word she once heard me say). Intrigued, she finally found a bookstore with that book and brought it home to me as a gift. At first I was not grateful since I faithfully avoid UFO books. For, though they often mentioned things with extraordinary flight behaviour, they invariably lacked any evidence at all. But this book had fairly compelling photos and described significant testing and analysis of the photos and metal samples allegedly associated these craft. So, I read this book.

It was a Swiss farmer's story about contacts with human-like beings over several years in the 1975 time-period. Mentioned, were fairly frequent visits by these allegedly space-faring beings whose "beam-ships" travelled 500 light years of distance (more than 1,000 trillion kilometers) between their home planet in the Pleiades star cluster and Earth. They made this long trip in only 7 hours, which would require average speeds more than 100 million, billion times the speed-of-light. I assumed the farmer may have been told all this by some person versed in science fiction. But the very short travel times claimed for the ships implied much higher acceleration and speed than one might ever expect. For these ships spent almost all their flight-time traveling an insignificant part of their total flight distance. Thus, stupendous acceleration to stupendous speed was required in only seconds of time. But instead of this seeming absurd flight profile increasing my disbelief in the farmer's story, it made me more intrigued.

The alien's craft (called "beam-ships") accelerated swiftly to high speed in about 3.5 hours of earth-time by means of a "light-emitting drive" and a "tachyon-drive" (that hurled acted the ship over a 250 light-year distance toward its target star in only several seconds of Earth time) to extraordinary speed which would have had to be over a trillion times c. As one would expect, a beam-ship slowed at similar high rates in travelling the remaining 250 light-years of distance to the star in an additional 3.5 hours of ship and Earth time. I assumed prevention of prohibitive ship stresses during very high acceleration would require something like "spacetime-warping" or "polarizing of quantum vacuum" by ship drives, to prevent ship mass and time dilation as light-speed is approached and surpassed. The farmer was told time passed at the same rate on both ship and Earth. This would be consistent with the ship drives preventing relativistic time-slowing or length shortening or mass enlarging, compared to when it was at rest on Earth.

The visitors said their tachyon drive took only seconds of time to achieve a "hyperspace-condition" where both space and time ceased for a moment before the ship accelerated in several seconds to peak speeds I estimated at more than a billion times faster than c. Interestingly, "space and time ceasing to exist" occurred at the bottom of my $k\tau$ coordinate in this paper's first figure. And a ship could, indeed, quickly accelerate in seconds to speeds billions of times light speed in planes of existence only slightly inclined from the vertical x-$k\tau$ plane embedded in my deeper x-ct-$k\tau$ realm. Finally, if beam-ships and tachyon drives existed, they would have to emit EM fields that favorably couple with those that give rise to gravity and inertia. So such systems, would have to emanate extraordinary fields that do extraordinary things - such as indicated Figure 3.

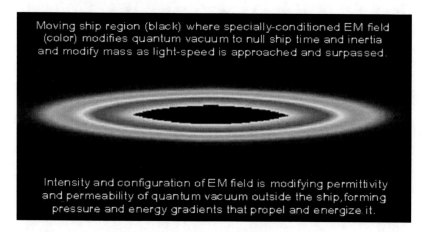

Figure 3. Ship path in "hyperspace.

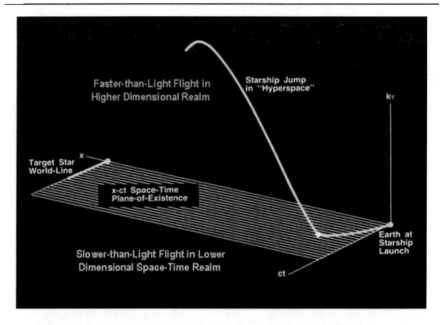

Figure 4. Ship path in 'hyperspace'.

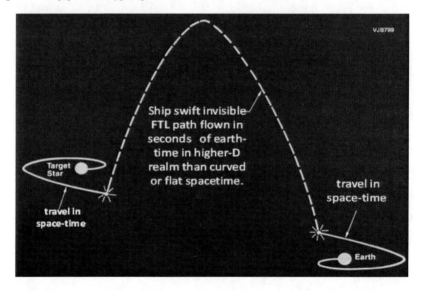

Figure 5. Short STL ship travel in space-time, and long FTL travel over 500 light years of distance. This trajectory idea was reported in [3] with no performance cited and no mention of UFOs.

Using my previously derived STL, FTL representations, I could crudely approximate a beam-ship trajectory between the visitor planet and Earth. Figure 4. shows ship path in 'hyperspace'. As shown in Figure 4, Figure 5 shows STL flight occurring on (or close to) the x-ct space-time plane until light-speed c is reached, and then after c is exceeded, FTL ship flight transitions to a near-vertical plane near to the x-ik space-tau one.

7.2. FTL Travel by Warping Spacetime and Modifying Matter with Special EM Fields

Most FTL drives for warping space-time metric or "polarizing zero-point vacuum" have been found wanting (Figure 6). For their EM fields couple weakly with the fields that give rise to gravity and inertia. Thus needed energies for warping or polarizing are enormous. But needed EM energies may not be prohibitive for conditioned EM fields like those being developed by T.W. Barrett [3]. His electric E and magnetic B fields, are described by higher tensor mathematics and they create more complex and subtle actions from 3 couplings between E and B and A tensor fields. These SU(2) radiation fields possess the same, Lie group symmetry as the SU(2) matter fields of the weak nuclear force. And the three A, B and E-field couplings are somewhat similar to the three vector bosons of the weak force. Shown below in Figure 7 is the Expanded Maxwell Equations for SU(2) EM fields.

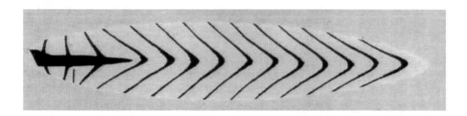

Figure 6. Special EM field.

Maxwell Tensor Equations For SU(2) EM Fields

$$\nabla \bullet E = J_0 - iq(A \bullet E - E \bullet A)$$

$$\frac{\partial E}{\partial t} - \nabla \times B - J + iq[A_0, E] - iq(A \times B - B \times A) = 0$$

$$\nabla \bullet B + iq(A \bullet B - B \bullet A) = 0$$

$$\nabla \times E + \frac{\partial B}{\partial t} + iq[A_0, B] = iq(A \times E - E \times A) = 0$$

Lorentz Force for U(1) and SU(2) EM Fields

U(1) Lorentz Force	$\mathscr{F} = eE + ev \times B = e\left(-\frac{\partial A}{\partial t} - \nabla\phi\right)$ $+ ev \times \left((\nabla \times A)\right)$
SU(2) Lorentz Force	$\mathscr{F} = eE + ev \times B = e\left(-(\nabla \times A) - \frac{\partial A}{\partial t} - \nabla\phi\right)$ $+ ev \times \left((\nabla \times A) - \frac{\partial A}{\partial t} - \nabla\phi\right)$

Figure 7. Maxwell Equation for SU(2) EM fields and different Lorentz force acting on charged matter moving in U(1) and SU(2) EM fields.

Figure 8 is ordinary U(1) EM energy in a wave guide transformed into SU(2) EM wave energy by phase and then polarization modulation, resulting in conditioned EM wave energy, rotating through every possible polarization in only femto-seconds of time.

Rapid Transit by Field Propulsion to Distant Stars 169

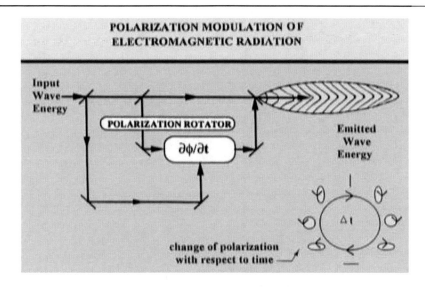

Figure 8. Polarization modulation of electromagnetic radiation.

As shown in Figures 9, 10, 11, rapid change in E-field magnitude and direction during 1 polarization modulation cycle during one wavelength of SU(2) beam travel. An Identical B-field variation occurs at 90 degrees phase offset.

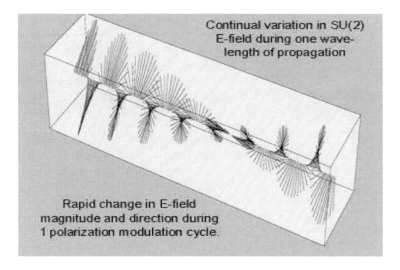

Figure 9. Rapid change in E-field direction and magnitude during 1 polarization-modulation cycle.

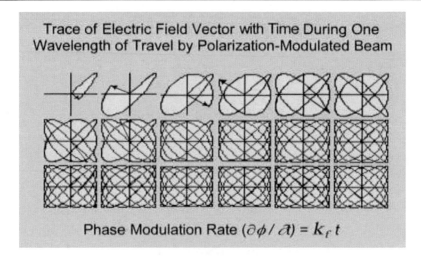

Figure 10. Trace of E-field magnitude and direction in vacuum during 1 polarization cycle.

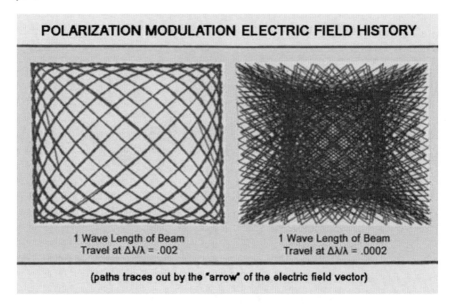

Figure 11. Shown, are many electric field paths traced over a spatial area during a very short EM beam travel distance for modest polarization modulation and many more traced out for more polarization.

What Can Give Rise to a Higher Dimensional Realm for Both STL and FTL Flight?

The UFO case didn't increase my belief in them, but it increased my conviction that FTL flight required a higher dimensional realm than curved space-time. I viewed such a higher realm as symbolizing a higher-condition than is seen and felt in a lower realm of space and time and energy-matter. And this led to the bizarre idea of a higher-D realm also being the result of a higher action – that of a specially-conditioned EM field radiated by a moving ship.

Region of Un-Labored Motion in a Higher Dimensional Realm than Ordinary Space-Time

The UFO case didn't increase my belief in UFOs, but it increased my conviction that reasonably unlabored FTL flight would require the added ship degrees of freedom and states of being that higher dimensionality would allow. Figure 12 shows, by the deeper volume of a 3-D realm that rises-above and descends below the area of a 2-D space-time plane of existence. This plane allows 1-D ship motion to the left or right of the time coordinate ct. The 2 quadrants in either direction from the $y = ct$ and $y = -ct$ time coordinate displays ship motions from zero to speeds as high as c and slowing to 0 speed. Movement in the -ct direction of the y coordinate corresponds to backwards-in-time travel into the past and, possibly, negative entropy which would create order out of chaos - not inevitable decay. But science almost universally forbids negative time or negative entropy within the plane shown below.

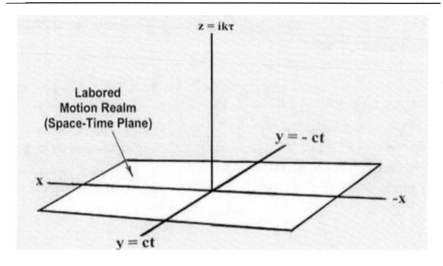

Figure 12. Shown is our familiar material, physical seen world symbolized as a 2-D x-ct space-time plane-of-existence embedded in the volume of a deeper, higher x-ct-ikτ realm rising above and going below the x-ct plane. Not shown are Special Relativity forbidden zones where space travel dx must not exceed time travel cdt and backwards time travel is forbidden in the –cdt direction.

On the top Figure 13 of the unlabored acceleration from 0 to as high as infinite speed is shown taking place on the surface of 1 quadrant of a cone. Here, momentum-conserving ship acceleration take place on this conic surface if the magnitude of the ship's momentum vector $[p_r + ip_i]$ is maintained at its initial magnitude m_0V_0 and if ratio of ship mass-rate-of-change to ship velocity rate-of-change $[(dm/dt)/(\mathbf{dV/dt})]$ equals the ever-changing ship mass-to-velocity ratio m/V during its acceleration. And during this acceleration, ship energetics is maintained at its initial E_0/c magnitude by the circumferential curvature of the conic. Figure 13 shows the deeper realm in the surface of a cone rising above the x-ct space-time realm of existence, and it can be shown it contain all possible worldliness of unlabored ship paths.

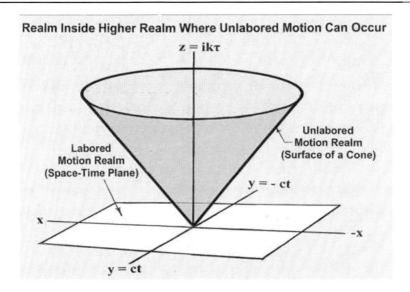

Figure 13. Conical surface where unlabored acceleration can be achieved from zero to infinite speed.

The conic surface extends upward to infinity, containing straight and curved worldliness for coasting and accelerating flight, for all possible unlabored ship accelerations. Using polar coordinates: $r = [x^2 + y^2 + z^2]^{1/2}$ and $\psi = \tan^{-1} V/c$ are sometimes more convenient than Cartesian ones for displaying unlabored worldliness of momentum and energetics-conserving ship space-time-tau paths upon the conic surface. On it, a trip to a distant star would be an ascending spiral around the rising z-coordinate (somewhat like the rising-circling-soaring flight of a bird. And such ascending flight in the vertical $ik\tau$ direction would be one-half of a revolution around the cone. By contrast, return to Earth from the star would require another ascending spiral upon the cone's surface to maximum height, followed by another descending spiral and another half revolution around the cone,- to thereby bring the ship back to Earth.

Such extraordinary STL-FTL flight possibilities, are, unfortunately, based on simple models that depend on an unusual thing. But they also reveal a possibility for another unusual thing. The unusual thing that must happen for continual ascending and descending flight on one-way or round-trip journeys to distant stars is the traversal of something like a singularity if near-infinite ship speed relative to Earth can be reached. Here, the above conic surface becomes tangent to one representing an SR forbidden region - as shown on a previous page. It is not clear that a tangency point would be a true barrier to a system moving at near-infinite speed and such a singularity may not exist in an actual higher realm of many more than three dimensions. But this remains acritical issue. Another thing is that "arrow of time" reversal occurs after reaching and slowing from near-infinite speed. Time reversal in a higher-D realm would not necessarily involve time-travel into the past, but if arrow-ot-time reversal is manifested in entropy reversal, its conceivable that renewal (not deterioration) of human cells might occur during unlabored ship deceleration in higher-D realms. This elimination of net human aging during unlabored flight may of course not happen in actual higher-D realms. But it is a bizarre possibility if higher-D realms for field-propelled spaceflight actually exist.

What Can Give Rise to a Higher Dimensional Realm for Both STL and FTL Flight?

A higher-D realm may manifest a deeper reality, wherein systems are endowed with an added degree of spatial or temporal freedom than is available for ordinary STL flight by ordinary energy-matter manipulation. And if these endowed systems can conserve their energetics and momenta while undergoing acceleration to different speed with neither energy or force expenditure, this leads to the idea shown below of a higher- realm being a higher-condition brought about by extraordinary actions by extraordinary fields radiated from moving ships (Figure 14).

Rapid Transit by Field Propulsion to Distant Stars 175

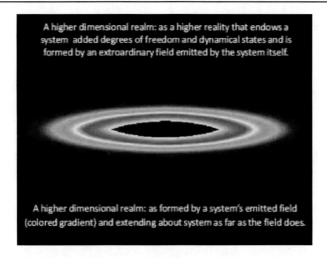

Figure 14. Higher realm formed by the extraordinary field emitted by a moving system like a ship.

Figure 15 shows a FTL path flown by warping space-time-tau metric in a higher realm. One way to accomplish this might be specially-conditioned EM field energy deposition into the quantum vacuum. And the associated interaction may enable warping realm's metric into the complex topology needed for energy-momentum conservation during ship's unlabored acceleration.

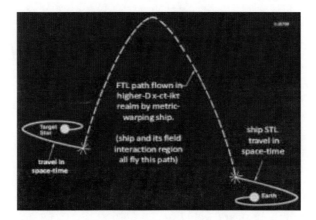

Figure 15. Idea of a higher-D realm for STL and FTL travel being part of a higher reality accessible by more enlightened reality is by the emitting of conditioned EM radiation fields that manifest higher symmetry and configuration.

Representation of System Kinematics and Dynamics by Complex Variable

Except for distance x, all system quantities in a higher-D x-ct-ikτ Minkowski Space were represented by complex variables of form: $q = q_r + iq_i$ with "real" and "imaginary" values as is shown in Figure 16. Here, the real and imaginary parts of a system's complex quantity gave it a directional, vector-like nature instead of only having a scalar nature in a lower-D x-ct space-time Minkowski Space.

Special relativity for systems in higher-D x-ct-ikτ Minkowski Space is consistent with ordinary Special Relativity ib x-ct Minkowski Space, in that light-speed c is invariant in all reference frames. But here, V/c, which is dx/c(dt) or dx/dy, is dx/(dy + idz) where: $dz = k(d\tau)$ and dy + id x must be more than dx.

Quantities Described by Complex Variables in a Higher Dimensional x-ct-iz Realm

- *Distance in space :* x $x = x$
- *Passage of time :* t $t = t_r + i\,t_i$
- *Velocity :* V $V = V_r + i\,V_i$
- *Mass (Energy/c^2) :* m $m = m_r + i\,m_i$
- *Momentum (mV) :* p $p = p_r + i\,p_i$
- *Energy: (mc^2)* E $E = E_r + i\,E_i$

Subscript : r = real part; i = imaginary part

Figure 16. Complex Quantities in higher dimensional realm.

Figure 17 shows that things get more complicated when system variables (like ship momentum state) involve multiplication of complex quantities like ship mass and velocity.

Rapid Transit by Field Propulsion to Distant Stars

Quantities Described by Complex Variables in a Higher Dimensional x-ct-iz Realm

- *Mass:* (E/c^2) m $m_r + i\, m_i$
- *Velocity:* V $V_r + i\, V_i$
- *Momentum* mV $(m_r + i\, m_i)(V_r + i\, V_i)$
- *Momentum* mV $(m_r V_r - m_i V_i) + i(m_i V_r + m_r V_i)$

Subscript: r = real part; i = imaginary part

Figure 17. More complex quantities in the higher dimensional realm.

Simple Minkowski Space for Approximating Kinematics of STL and FTL Flight

A 3-D Minkowski Space [4] with 1 spatial, temporal and extra dimension was constructed as in the following Figures 18, 19, 20, 21 to represent a higher-dimensional realm than a 2-D space-time.

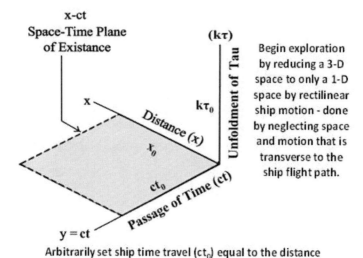

Arbitrarily set ship time travel (ct₀) equal to the distance light travels at its speed (c) during a ship's space travel dx₀.

Figure 18. Spacetime plane of existence.

Herman D. Froning, Jr.

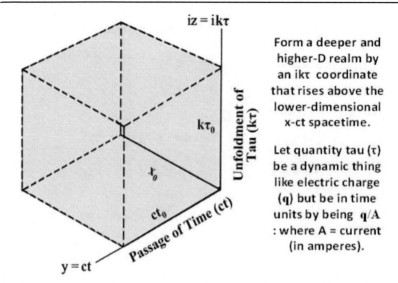

Figure 19. Higher realm than space-time.

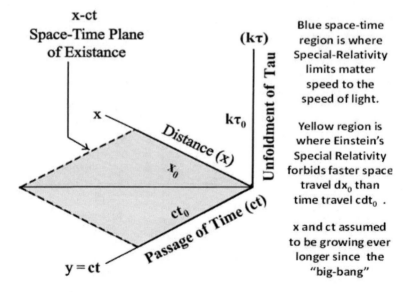

Figure 20. Slower-than-light plane of existence.

Rapid Transit by Field Propulsion to Distant Stars

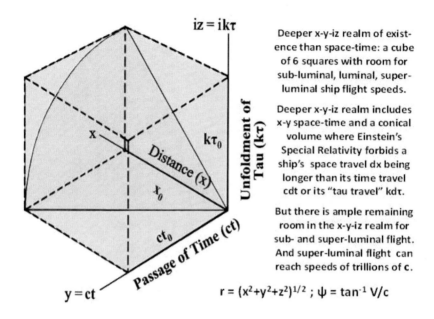

Deeper x-y-iz realm of existence than space-time: a cube of 6 squares with room for sub-luminal, luminal, super-luminal ship flight speeds.

Deeper x-y-iz realm includes x-y space-time and a conical volume where Einstein's Special Relativity forbids a ship's space travel dx being longer than its time travel cdt or its "tau travel" kdτ.

But there is ample remaining room in the x-y-iz realm for sub- and super-luminal flight. And super-luminal flight can reach speeds of trillions of c.

$r = (x^2+y^2+z^2)^{1/2}$; $\psi = \tan^{-1} V/c$

Figure 21. Slower and faster-than-light realm of existence.

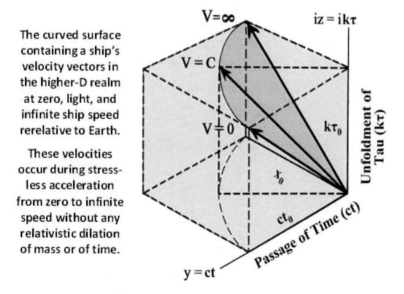

The curved surface containing a ship's velocity vectors in the higher-D realm at zero, light, and infinite ship speed rerelative to Earth.

These velocities occur during stress-less acceleration from zero to infinite speed without any relativistic dilation of mass or of time.

Figure 22. Unlabored acceleration from zero to infinite speed.

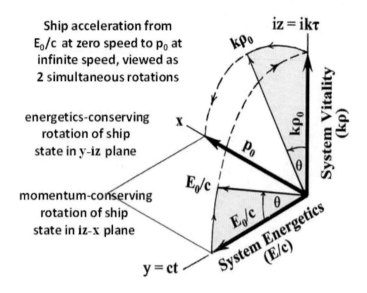

Figure 23. Unlabored rotation of ship state to infinite speed.

Figure 22 and Figure 23 are a conic surface in the higher-D realm that contains the velocity vectors of all moving systems that suffer no relativistic distortion at all. This is because all system world-lines on this surface are tangent to Special Relativity's conic region of forbidden flight. Thus no light barrier exists for any flight system that is upon this plane. And the next figure shows 1 way to achieve near-infinite-speed by 2 system state vector rotations in the higher realm.

Note: Conservation of system angular momentum not included in analysis.

A final effort was a simulation by R.L. Roach of exceeding light-speed by beamed EM energy emitted into quantum vacuum from an accelerating wing-shaped ship. Shown is the fairly complex computational grid needed to model the ship, its EM beams, and affected vacuum.

Requirements for Unlabored Momentum and Energy-Conserving STL and FTL Flight

Newtonian physics conservation laws decrees that a system's acceleration to different speed requires energy and force expenditure by it to maintain (conserve) the system's energy and momentum at their values prior to the system's acceleration. By obeying the momentum-conservation law, a system can undergo force and stress-free acceleration in space-time by reducing its mass such that its momentum mV remains conserved. But system internal energy mc^2 is not conserved unless the energy lost during its acceleration is replaced by energy from outside it. But an added system degree-of-freedom and dynamic state in a higher-D x-ct-ikτ realm may make energetics-conserving system acceleration a possibility. Figure 24 shows some requirements for momentum and energy-conserving flight in the x-ct and x-cf-ik realms.

Figure 24 are examples of unlabored energy and momentum-conserving ship acceleration in the higher x-ct-ikτ realm. The top one shows conservation of ship energetics E_0/c while system vigor develops to value $k\rho_0$. It also shows conserving of the magnitude of the system's momenta E/c + ikρ during acceleration from zero to infinite speed relative to Earth. The bottom figure shows another unlabored ship trajectory from zero to infinite speed in the higher realm. It is energy-conserving transformation of a ship's initial energetics E_0/c to $ik\rho_0$ in the higher realm's space-less ct-ikτ plane; then momenta-conserving transformation of the ship's initial momenta from $ik\rho_0$ to p_0 in the time-less x-ikτ plane of the higher realm.

Of special concern for ship's navigation-guidance-control is the prevention of excessive ship stresses developing in any x-ct space-time part of its higher-dimensional x-ct-ikτ realm. This requires dp/dt = d(mV)/dt = V(dm/dt) + m(dV/dt) during. Combining this relation with the need for ship mV be maintained at its initial momentum m_0V_0 during ship acceleration, results in: **dm/dt = [dV/dt] [(dm)(m₀V₀)(m/V²)]** in order to maintain force-free and stress-free ship acceleration. So, if constant ship acceleration is desired, ship guidance and control systems must diminish mass-warping rate as the ship's mass lessens and its speed increases.

182   Herman D. Froning, Jr.

Requirements for Unlabored Energetics and Momenta-Conserving Acceleration within a Higher-Dimensional x-ct-ikτ Realm

- Ship momenta $(p_r + ip_i)$ conserved in higher realm and momentum p_r conserved in x-ct space-time $(dp_r/dt_r = 0)$

- Ship energetics $(E = E_r + iE_i)$ conserved in higher realm $(dE/dt = 0)$ as energy E_r lessens in x-ct space-times.

- Stress-less ship acceleration by increasing its kτ state in higher realm's space-less ct-ikτ plane - where $dV_r/dt_r = 0$,

- Stress-less acceleration by increasing system kτ state in higher realm's time-less ikτ-x plane - where $dV_r/dt_r = 0$

Figure 24. Requirements for unlabored system acceleration.

Alqubierre Warp Drive for Warping of Curved Spacetime Metric

As has been mentioned, Miguel Alqubierre has defined a spacetime-warping solution of General Relativity (GR) that allows ship acceleration from STL to FIL speed. Figure 25 is a geometric representation of his solution. Seen is vertical warping of a 2-D gravitational metric that is embedded in a higher 3-D "empty" space. Curved spacetime metric is moderately warped in the ship to prevent un-acceptable stress developing in it during acceleration and more extreme upward-downward warping around the ship periphery as shown in Figure 26.

Rapid Transit by Field Propulsion to Distant Stars

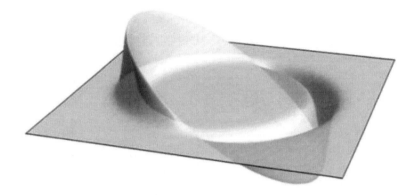

Figure 25. This presentation above and some of my previous ones are similar in that both show actions in a 3-D realm where lower-D spacetime metric is embedded. So, just as this representation above shows warping of embedded, lower-D metric within a higher 3-D realm, some of this Section's previous figures show much less-labored ship trajectories that require less warping in that higher 3-D realm.

> Less warping of space-time metric is needed if favorable interaction by ship-emitted EM fields can greatly reduce or reverse the resistance of those fields that give rise to the accelerating ship's mass and inertia.

> So, a higher-D realm that significantly conserves ship momenta $(p_r + ip_i)$ and energtics $(E_r + iE_i)$ during acceleration, reduces needed amounts of space-time warping done by the ship.

Figure 26. Also shown is vertical warping of spacetime metric to cause ship impulsion. Alqubiere's work stimulated much added work by many others for many years. Unfortunately, most of this additional work was discouraging. For it found that extremely weak coupling between gravity and ordinary EM fields required the generation of stupendous amounts of EM field energy to warp space-time metric into the Alqubierre's topology below that enables faster-than-light travel.

A final effort [5] was a preliminary simulation of ship transition from STL to STL state by EM energy beamed into the quantum vacuum from an accelerating wing-like ship. This was done with major help by Dr. Robert Roach at Georgia Tech University. Figure 27 shows the fairly complex computational grid needed to model the ship and EM energy deposited into vacuum.

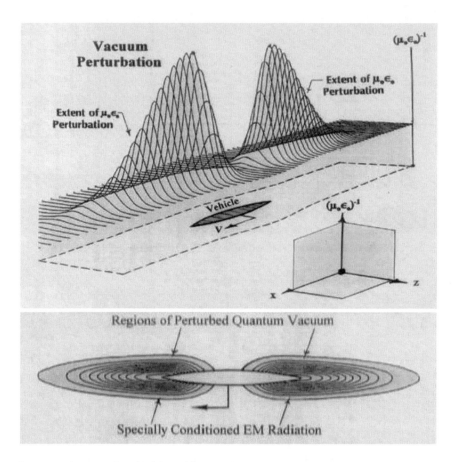

Figure 27. Computational grid used by R.L. Roach to model beamed-energy emitted from a 2-D wing.

Figure 28 is favorable distorting of the ship's surrounding vacuum by EM discharges that reduce its electrical permittivity and magnetic permeability in front of and behind the accelerating ship at 0.99c flight speed relative to Earth. This work by Roach used computational fluid dynamics (CFD) by assuming some degree of similarity in fluid and electromagnetic energy transfer in compressible flight mediums like air and quantum vacuum.

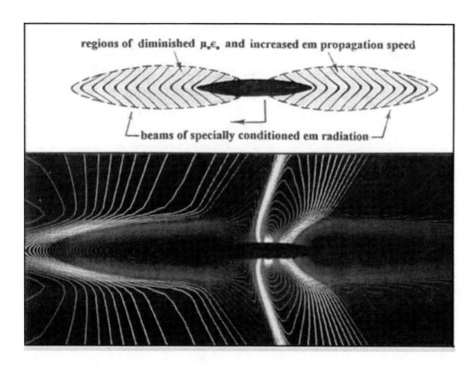

Figure 28. Reversal of vacuum resistance to ship acceleration by favorable redistribution of vacuum zero-point energy and pressure by beamed energy deposition into vacuum from ship moving at .99 c.

Shown is the discharge causing significant vacuum modification and radiation pressure gradients that resulted in ship impulsion (rather than resistance). Roach's work revealed some encouraging possibilities for reducing or reversing vacuum resistance to vehicle motion by perturbing it with beamed energy. This also created favorable zero-point energy gradients for possible energy extraction. But the simulation work also revealed staggering difficulties and many unknowns and lack of much knowledge for proper simulation of field-propelled flight.

Bob roach and I were jumping with joy when his last run indicated that flight-assisting pressures (rather than resisting ones) had pushed our vehicle through the light barrier. I also patted him on the back for being the first to fly a wing through the sound barrier by actions and reactions of fields (rather than by combustion and expulsion of mass). But we didn't gloat long because we knew a flight-size version of our field propulsion system would expend far more energy than a rocket or jet engine today. And we didn't brag, because we may have been successful because of our ignorance – because we left out some friction loss or exotic dissipation term like Chrenkov radiation. And, since we were assuming emitted EM fields were also nulling relativistic dilation and maintaining constant ship momentum by reducing ship mass with increasing speed, we had only addressed the external resistance from our ship's disturbed flight medium that surrounded it. However, our work, defined in somewhat more detail below, revealed intense EM field energy deposition into a medium (like air or the quantum vacuum) might annihilate much medium resistance to high-speed flight. And since the more formidable problem of developing EM fields to do all this went un-addressed, our simulation probably did nothing to reduce the stupendous challenge and uncertainty of faster-than-light flight. Yet, it left us with a good feeling and a glimmer of hope that the seemingly impossible dream of faster-than-light flight might conceivably come to pass.

7.3. A Proposed Initiative for Advancing Field-Propulsion

A lot happened between 1953 and 1983 when the most rapid advances in human space-flight occurred. Flight progress seemed to diminish shortly thereafter. But aerospace conferences and talking about space-flight seemed to increase. And some of the many talks about planned government initiatives to advance spaceflight seemed depressing and somewhat boring – even though they were expensive and lasted far into the future. But none seemed bold enough to revolutionize spaceflight: to reduce its cost enough to make it a commercial-economic opportunity instead of continual taxpayer burden. Also, NASA was rather poor in controlling its spaceflight costs. Space Shuttle costs were then about 40 times more than promised and the Space Station costs were exceeding estimates by factors of 10. So a NASA estimate of 450 billion taxpayer dollars for a planned Mars Expedition with chemical rockets was increased by cost experts to over a trillion US dollars.

Reasons and goals for spaceflight initiatives also seemed somewhat mundane, though they could always claim good things like "increasing youth interest in science" and "maintaining U.S. pre-eminence in space". So, after listening to a great many plans for future spaceflight, I decided to prepare one of my own. And I began by reviewing some past space flight advances during the first century of flight. They are shown in Figure 29.

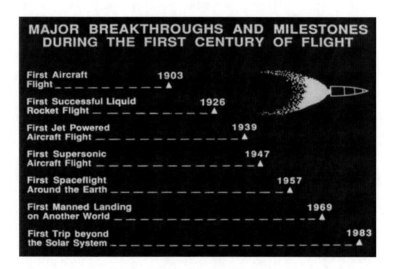

Figure 29. Major propulsion and flight advances during the world's first 80 years of powered flight.

I attempted to extrapolate past advances in flight speed with time during the first 80 years of flight on into the future (the next 80 years of flight). I couldn't find enough data for a valid one. But extrapolating past advances in flight distance was found more possible and showed that the most nearby star could be reached before the present twentieth century ends. So I put together an estimate of future nuclear fusion and field power and propulsion developments that might be somewhat comparable to chemical rocket engine and jet engine and nuclear fission rocket engine developments accomplished in the past. And these future developments, shown on Figure 30, the next page, enable field-propelled Mars flight by 2050 and star flight by 2100.

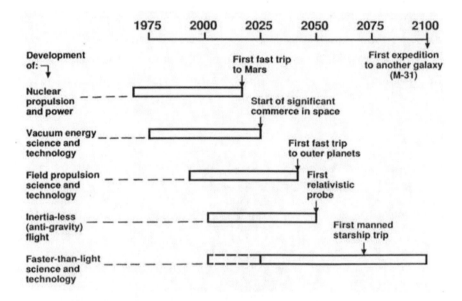

Figure 30. Field-propelled Mars flight by 2050 and star flight by 2100.

I called my Initiative an "Interstellar Initiative for Future Flight" and envisioned it as a fairly bold theoretical and experimental activity that would advance the science and technology needed for the development of nearly-propellant-less power and propulsion by actions of fields - not combustion of matter. So, it might better have been called "An Initiative for Field Power and Propulsion Development". I didn't envision it as developing, building or operating space ships, but rather developing field power and propulsion science and technology through "proof-of-principle" or "prototype system" demonstrations. By contrast, full-scale development and flight testing would be done by existing government-industry organizations. However, technology advances by the initiative would enable vehicles as shown in Figure 31.

Rapid Transit by Field Propulsion to Distant Stars

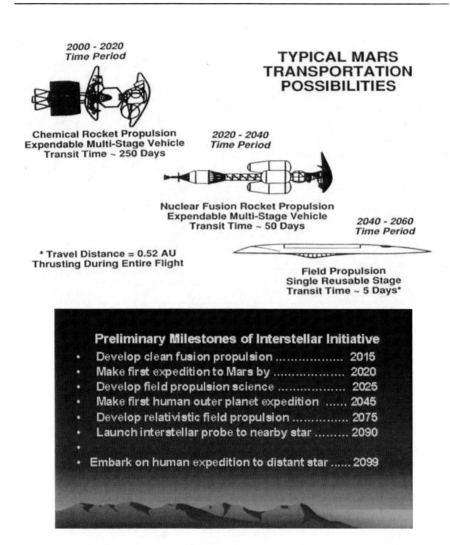

Figure 31. Preliminary technical and mission goals for future fusion and field-propulsion advancements.

I documented my work in a paper [1] and was privileged to present it at many forums around the world in the 1990's. My many presentations stimulated interest from scientists, engineers and university students for about 5 years. But advancement remained slow as technology leaders decided on safe slow research - not the bolder-faster-riskier things that were done at a higher tempo when there was the urgency of peril from a perceived threat.

A paper [2] by Michael Papagiannis of Boston University gives insight into issues associated with development of space-faring civilizations and their survival or fall. He began with their industrial phase, which usually began with un-checked growth, followed by limits to growth by overpopulation, resource depletion and environmental degradation. He said that physical strength and procreation ability, important in early evolution, must finally be superseded by collective wisdom and ethical development to control population and materialistic expansion. But this would require unselfish societies who had conquered belligerence and lost interest in wealth. And he argued that only moral and ethical societies survive long enough to develop the needed wisdom, science and technology for extraordinary things like interstellar flight.

If Papagiannis is right, things like interstellar initiatives and propellant-less field propulsion will never come to pass if significant advances in human wisdom, ethics and morality are not accomplished. So, what motivation can cause Earth societies or civilizations to advance in wisdom-ethics-morality- in addition to science and technical innovation? Obviously my past attempt to motivate humanity to accomplish an extraordinary thing "Star flight before the end of this Century" was a failure. This, in part, was because there seemed no real urgency in accomplishing it. But now there seems an urgency in accomplishing an extraordinary thing before this century ends. And this thing may require propellant-less power and propulsion.

This extraordinary need is not departing for a star before this century ends. We have many more centuries before we have to do this (when our Sun's nuclear energy that gives us light and heat begins to fail). But science now assures us that Earth's temperature must not rise more than several degrees during its next 85 years. And best estimates indicate combustion of matter must be cut in half by about 2050 and must almost end by the time this century ends.

Figure 32 shows oxygen-hydrogen propellant reductions I estimated for affordable, cost-effective interplanetary flight to Mars in 2050, and someone

of many science estimates of needed reductions in gas-oil-coal burning by 2050 to prevent excessive global warming by 2100.

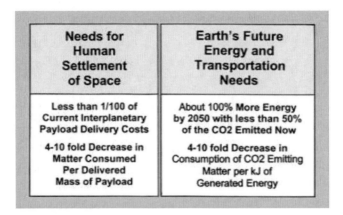

Figure 32. Safe, economical interplanetary flight in 2050 and safe global temperatures on Earth in 2050 require enormous reductions in consumption of matter.

Interestingly, my Interstellar initiative goal of embarking on the first star-flight by the end of this twentieth century coincides with the climate-science goal of limiting Earth's temperature to a rise of no more than about three degrees more than on the same date when this Century ends. And it is significant that achieving either goal surely require almost total abstinence from the combustion (heating) of the same kinds of matter used for our fuel today. And. as mentioned by Papagannis, this may require a more wise, moral and ethical world - one that has arisen significantly above selfishness and greed - and one that relies on fields, not on matter.

REFERENCES

[1] Froning, H.D. Jr., "A Metaphysical Interpretation of Tachyons", *Speculations in Science and Technology*, Elsevier, Sequoia S.A. Lausanne Switzerland, 1981, 1982.

[2] Stevens, W.C., *UFO, Contact from the Pleiades*, ISBN 0-9608558-2-3, *UFU Photo Archives*, Tuscon Arizona, ISBN 9608558, 1982.

[3] Froning, H. D. Jr., "Propulsion Requirements for Rapid Transits to Distant Stars", *Journal of the British Interplanetary Society*, Vol. 36, No. 5, May,1985.

[4] Froning, H.D. Jr., "Swift, Unlabored Vehicle Acceleration by Momentum and Energy-Conserving Paths in Higher Dimensions", *47th AIAA/ASME/SAE/ASEE Joint Propulsion Conference and Exhibit*, San Diego California, August, 2011.

[5] Froning, H.D. Jr., "Preliminary Simulations of Vehicle Interactions with the Zero-Point Quantum Vacuum by Fluid-Dynamic Approximations", *AIAA 2000-3478, 36th AIAA/ASME/SAE/ASME Joint Propulsion Conference and Exhibit*, Huntsville Alabama, July, 2000.

[6] Froning, H.D., "An Interstellar Exploration Initiative for Future Flight", MDC 91 H7041, Mc Donnell Douglas Space Systems Company, Presented at the *28th Space Congress*, Cocoa Beach, Florida, April, 1991.

[7] Papagainnis, M.J., "*Natural Selection of Stellar Civilizations by the Limits of Growth*", Astronomy Department, Boston University, 1984, <html:file:1984ORAS-25-309>

Chapter 8

HYPER-SPACE NAVIGATION

Yoshinari Minami

8.1. INTRODUCTION

As is well known in astronomy, sixty-three stellar systems and other eight hundred fourteen stellar systems exist respectively within the range of 18 and 50 light years from our Solar System. For instance, Alpha Centauri is the nearest star from Earth, and the star Sirius, which is the seventh nearest star, is 8.7 light years from Earth, while the Pleiades star cluster is 410 light years from us. According to Einstein's Special Relativity, sending a starship to a stellar system at a distance longer than several hundred light years would ask for an extremely long time even if the starship would travel at the speed of light. For instance, assuming that the starship is traveling to the Pleiades star cluster at a speed of $0.99999c$, it will arrive at the Pleiades 1.8 years later and, in the case of immediately starting of the return travel, it would be back to Earth 3.6 years after leaving for the Pleiades. But this would be just for the clocks of the astronauts onboard the starship for that mission. For people on Earth, the whole time period would be 820 years, with paradoxical consequences as to the feasibility of a mission such as this. The first solution of the above-stated problem is to obtain a breakthrough in propulsion science. However, from the standpoint of propulsion theory using not only momentum thrust but also pressure thrust, there is no propulsion theory which exceeds the speed of light, even if we use the field propulsion theory. Accordingly, the propulsion theory alone is not enough to establish the reality of interstellar travel, thereby requiring a navigation theory as a secondary solution.

Concerning interstellar travel or intergalactic exploration, the method using a wormhole is well known; relying on space warps, such as for instance Wheeler-Planck Wormholes, Kerr metric, Schwarzschild metric, Morris-Thorne Field-Supported Wormhole based on the solutions of equations of General Relativity [1]. However, since the size of wormhole is smaller than the atom (10^{-15}m), i.e., $\sim 10^{-35}$m and moreover the size is predicted to fluctuate theoretically due to instabilities, space flight through the wormhole is technically difficult and it is unknown where to go and how to return. Additionally, since the solution of wormhole includes a singularity, this navigation method theoretically includes fundamental problems. The search for a consistent quantum theory of gravity and the quest for a unification of gravity with other forces (strong, weak, and electromagnetic interactions) have both led to a renewed interest in theories with extra spatial dimensions. Theories that have been formulated with extra dimensions include Kaluza-Klein theory, supergravity theory, superstring theory, M theory, and D-brane theory related superstring. For instance, superstring theory is formulated in 10 or 26 dimensions (6 or 22 extra spatial dimensions). These extra spatial dimensions must be hidden, and are assumed to be unseen because they are compact and small, presumably with typical dimensions of the order of the Planck length ($\sim 10^{-35}$m). The navigation method of utilizing extra dimensions (even if they are compactified) has also a theoretical problem as well as using a wormhole.

On the one hand, there exists another interstellar navigation theory. Froning showed the rapid starship transit to a distant star (i.e., Instantaneous Travel) using the method of "jumping" over so-called time and space [2, 3] (Figure 1). The detail is explained in Chapter 7 in this book.

This idea is based on "tachyons" which could go faster than light (FTL). And some viewed tachyons as faster-than-light solutions of Einstein's

Special Relativity (SR). For, SR's Lorentz contraction factor $[(1-(V/c)^2]^{1/2}$ has real values only for speed V slower than light (STL). But, if multiplied by $i = [-1]^{1/2}$ (imaginary unit), this factor becomes $[(V/c)^2-1]^{1/2}$ which has real values only for FTL speed. Figure 1 shows the plane of existence of ordinary x-ct space-time. This was with a vertical coordinate ikτ, which is orthogonal to those which describe space travel (x) and time travel (ct) of slower-than-light tachyons on an x-ct space-time plane of existence. And this x-ct plane is seen to be embedded in the volume of the higher-dimensional x-ct-ikτ realm rising above it.

In addition to this invaluable concept, Minami studied its own navigation theory independently and proposed the Hyper-Space navigation theory using a space-time featuring an imaginary time [4, 5, 6, 7, 8, 9] (Figure 2). Hyper-Space navigation theory using a space-time featuring an imaginary time offers a great promise to develop practical interstellar exploration. Although it seems to be similar in using imaginary time, it is based on a totally different theory from Tachyon. This proposed navigation theory is based on Special Relativity (not on General Relativity), that allows interstellar travel to the farthest star systems to be realized; and removes the present theoretical limitations to interstellar travel that arises from the extremely long time needed (the time paradox) according to Special Relativity.

By the way, imaginary time is a difficult concept to grasp, and it is probably much difficulty that has caused most problems. How can imaginary time have anything to do with the real universe? Stephen Hawking has been working at developing equations that would tell us just what did happen when time began. The concept of imaginary time is related to the origin and fate of the universe. His theories use such concepts as imaginary time and singularities to unite relativity and quantum physics [10, 11].

As to Feynman's sum over histories, to avoid some technical difficulties, one must use imaginary time. In real time, the universe has a beginning and an end at singularities that form a boundary to space-time at which the laws of science break down. But in imaginary time, there are no singularities and boundaries. This might suggest that the so-called imaginary time is really more basic.

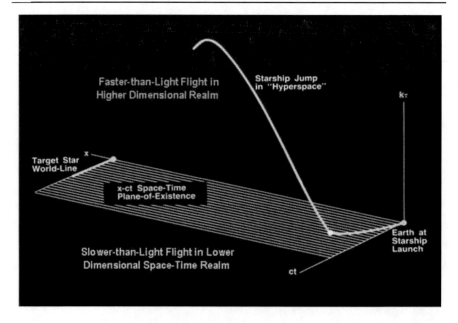

Figure 1. Trajectory faster-than-light flight (H.D. Froning).

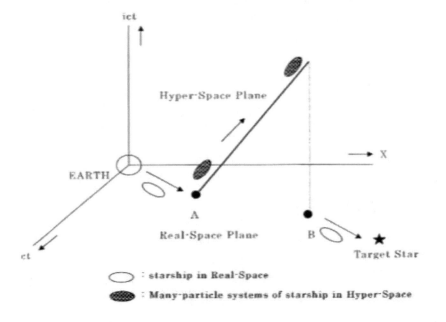

Figure 2. Interstellar travel to the star.

The practical interstellar travel combines propulsion theory with navigation theory. In the following sections, Hyper-Space navigation theory is described in detail.

8.2. THREE WAYS TO THE INTERSTELLAR TRAVEL

Three methods are considered to reach the star rapidly. The basic principle is the following equation which is known to every one:

$$L_{star} = V_{starship} \times t$$

where, L_{star} is the distance to star, $V_{starship}$ is the speed of starship, t is the time.

The distance to a stellar system "L_{star}" is enormous. An extremely long time is required, even if the starship would travel at the speed of light "c". To reach the star rapidly, three parameters, such as "speed", "distance" and "time", shall be controlled.

1) < **Change speed** > $L_{star} = (nc) \times t$

where, "nc" is n-fold increase in speed of light "c". Here, n is real number greater than 1.

There is no propulsion theory exceeds the speed of light, moreover, Special Relativity restricts the maximum speed to the speed of light; therefore this method is impossible.

2) < **Change distance** > $\dfrac{L_{star}}{n} = c \times t$

The so-called "wormhole" is utilized. By using wormhole, shorten the distance as $L_{star}/n \approx$ a few meters, as shown in Figure 3. For example, one meter in a wormhole corresponds to a few light years in actual space.

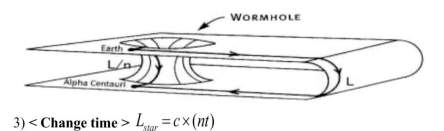

3) < **Change time** > $L_{star} = c \times (nt)$

Figure 3. A wormhole creates a shortcut from Earth to Alpha Centauri.

The time "t" in an imaginary time hole is equivalent time of n-fold time in actual space, as shown in Figure 4.

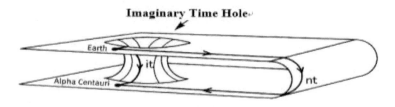

Figure 4. An Imaginary Time Hole creates a shortcut from Earth to Alpha Centauri.

For example, one second in an imaginary time hole corresponds to one million seconds in actual space.

Subsequently, interstellar travel through the imaginary time hole is described as the following.

8.3. HYPER-SPACE NAVIGATION SYSTEM

Properties of Flat Space

In general, the property of space is characterized by a metric tensor that defines the distance between two points. Here space is divided into two types. Actual physical space that we live in is a Minkowski space, and the world is limited by Special Relativity. It is defined as "Real-Space". Here as a hypothesis, an invariant distance for the time component of Minkowski metric reversal is demanded. This is not a mere time reversal. It is defined as "Hyper-Space". The invariance is identical with the symmetries. Symmetries in nature play many important roles in physics. From this hypothesis, the following arises: the properties of the imaginary time (x^0 = ict; i^2 = -1) are required as a necessary result in Hyper-Space. Here, "i" denotes the imaginary unit and "c" denotes the speed of light. The time "t" in Real-Space is changed to imaginary time "it" in Hyper-Space. However, the components of space coordinates (x,y,z) are the same real numbers as the Real-Space. From the above, it is seen that the real time (x^0 = ct) in Real-Space corresponds to the imaginary time (x^0 = ict) in Hyper-Space. That is, the following is obtained:

Real-Space: t (real number), x,y,z (real number);
Hyper-Space: it (imaginary number; i^2 = -1), x,y,z (real number).

The imaginary time direction is at right angles to real time. This arises from the symmetry principle on the time component of Minkowski metric reversal (see *Appendix D:* Properties of Hyper-Space).

Lorentz Transformation of Hyper-Space

Next, the Lorentz transformation of Hyper-Space corresponding to that of Real-Space is found.

Since the components of space coordinates (x, y, z) do not change between Real-Space and Hyper-Space, the velocity in Hyper-Space can be obtained by changing t → it:

$$V = \frac{dx}{dt} \rightarrow \frac{dx}{d(it)} = \frac{dx}{idt} = \frac{V}{i} = -iV \ . \tag{1}$$

The velocity becomes the imaginary velocity in Hyper-Space. Substituting "t→it, V→-iV" into the Lorentz transformation equations of Minkowski space formally gives:

< **Hyper-Space Lorentz transformation** >

$$x' = (x - Vt)\big/\sqrt{1 + (V/c)^2}, \quad t' = (t + Vx/c^2)\big/\sqrt{1 + (V/c)^2}$$
$$\Delta t' = \Delta t \sqrt{1 + (V/c)^2}, \quad \Delta L' = \Delta L \sqrt{1 + (V/c)^2}.$$
(2)

This result agrees with the results of detailed calculation. As a reference, the Lorentz transformation equations of Minkowski space, i.e., of Special Relativity, are shown below:

< **Real-Space Lorentz transformation: Special Relativity** >

$$x' = (x - Vt)\big/\sqrt{1 - (V/c)^2}, \quad t' = (t - Vx/c^2)\big/\sqrt{1 - (V/c)^2}$$
$$\Delta t' = \Delta t \sqrt{1 - (V/c)^2}, \quad \Delta L' = \Delta L \sqrt{1 - (V/c)^2}.$$
(3)

The main difference is that the Lorentz-Fitz Gerald contraction factor $\sqrt{1 - (V/c)^2}$ is changed to $\sqrt{1 + (V/c)^2}$.

Now, consider navigation with the help of both Lorentz transformations, especially the Lorentz contraction of time.

Figure 5 shows a transition of starship from Real-Space to Hyper-Space. In Figure 5, region I stands for Real-Space (Minkowski space). Consider two inertial coordinate systems, S and S'. S' moves relatively to S at the constant velocity of starship (V_S) along the x-axis. S' stands for the coordinate system of the starship and S stands for the rest coordinate system ($V_S = 0$) on the earth. Δt_{ERS} is the time of an observer on the earth, i.e., earth time, and $\Delta t'_{RS}$ is the time shown by a clock in the starship, i.e., starship time. Region II stands for Hyper-Space (Euclidean space). S' moves relatively to S at the constant velocity of starship (V_S) along the x-axis. S' stands for the coordinate system of the starship in Hyper-Space and S stands for the rest coordinate system ($V_S = 0$) in Hyper-Space. Δt_{EHS} is the time of an observer on the earth in Hyper-Space, i.e., the equivalent earth time, and $\Delta t'_{HS}$ is the time shown by a clock in the starship in Hyper-Space, i.e., the starship time. Now, the suffix."HS" denotes Hyper-Space and the suffix. "RS" denotes Real-Space.

Figure 5 also shows a linear mapping f: RS (Real-Space)→HS (Hyper-Space), that is, from a flat Minkowski space-time manifold to a flat imaginary space-time manifold. It is assumed that space is an infinite continuum [12]. There exists a 1-1 map f: RS→HS, $x^i|\to f(x^i)$ and a 1-1 inverse map f^{-1}: HS→RS, $f(x^i)|\to x^i$. The mapping is a bijection. These transformations will be local and smooth.

Now suppose that a starship accelerates in Real-Space and achieves a quasi-light velocity ($V_S \sim c$) and plunges into Hyper-Space by some new technical methods. Here, "plunges into Hyper-Space" means "transition from Real-Space to Hyper-Space", or is often said to be "Jump" or "pass through the barrier separating Real-Space from Hyper-Space".

In Real-Space, from Eq. (3),

$$\Delta t'_{RS} = \Delta t_{ERS} \sqrt{1-(V_S/c)^2} \ . \tag{4}$$

Eq. (4) is the so-called Lorentz contraction of time derived from Special Relativity. In Hyper-Space, the starship keeps the same velocity as the quasi-light velocity ($V_S \sim c$) just before plunging into Hyper-Space, i.e., $V_{S(HS)} = V_{S(RS)}$. Therefore, from Eq. (2),

$$\Delta t'_{HS} = \Delta t_{EHS} \sqrt{1+(V_S/c)^2} \ . \tag{5}$$

From Figure 5, after plunging into Hyper-Space, the starship keeps the quasi-light velocity and takes the S' coordinates. The elapsed time in the starship will be continuous. Considering the continuity of starship time between Real-Space and Hyper-Space, we get

$$\Delta t'_{RS} = \Delta t'_{HS} \ . \tag{6}$$

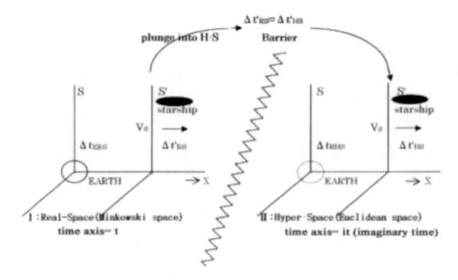

Figure 5. Transition from Real-Space to Hyper-Space.

Now Eq. (6) gives, from Eqs. (4) and (5),

$$\Delta t_{ERS} = \Delta t_{EHS} \left(\sqrt{1+(V_S/c)^2} \Big/ \sqrt{1-(V_S/c)^2} \right). \qquad (7)$$

Eq. (7) is the time transformation equation of earth time between Real-Space and Hyper-Space. From Eq. (7), when $V_S = 0$, we get

$$\Delta t_{ERS} = \Delta t_{EHS}. \qquad (8)$$

Namely, in the reference frame at rest, the elapsed time on the earth coincides with both Real-Space and Hyper-Space. However, as the velocity of starship approaches the velocity of light, the earth time between Real-Space and Hyper-Space becomes dissociated on a large scale.

Since an observer on the earth looks at the starship going at $V_S \sim c$ and loses sight of it as it plunges into Hyper-Space, it is observed that the starship keeps the same velocity and moves during the elapsed time Δt_{ERS}(at $V_S \sim c$) observed from the earth. Observer exists in S system. Therefore, the range of starship of an observer on the earth is given by

$$L = V_S \Delta t_{ERS} \approx c\Delta t_{ERS}. \tag{9}$$

For instance, in the case of $V_S = 0.999999999c$, from Eqs. (7) and (5), we get

$$\Delta t_{ERS} = \Delta t_{EHS} \times 31622, \quad \Delta t'_{HS} = \Delta t_{EHS} \times 1.4. \tag{10}$$

One second in Hyper-Space corresponds to 31,622 seconds in Real-Space. Similarly, one hour in Hyper-Space corresponds to 31,622 hours (3.6 years) in Real-Space.

While the starship takes a flight for 100 hours ($\Delta t'_{HS} = 100hr$; $V_S = 0.999999999c$) shown by a clock in the starship in Hyper-Space, 70 hours ($\Delta t_{EHS} = 70hr$; $V_S = 0$) have elapsed on the earth in Hyper-Space. Since this elapsed time on the earth in Hyper-Space is in the reference frame at rest, the time elapsed in it is the same as the time elapsed on the earth in Real-Space ($[\Delta t_{EHS}; V_S = 0] = [\Delta t_{ERS}; V_S = 0] = 70hr$). Therefore, there is not much difference between the elapsed time (70 hours) of an observer on the earth in Real-Space and the elapsed time (100 hours) of starship during Hyper-Space navigation. However, this elapsed time of 70 hours ($\Delta t_{EHS} = 70hr$; $V_S = 0$) on the earth in Hyper-Space becomes the elapsed time of 253 years ($\Delta t_{ERS} = 70 \times 31,622 = 2,213,540hr$; $V_S = 0.999999999c$) on the earth in Real-Space, because the starship flies at the velocity of $0.999999999c$. These 253 years represents the flight time of starship observed from the earth in Real-Space. Therefore, by plunging into Hyper-Space having the properties of imaginary time, from Eq. (9), the starship at a quasi-light velocity can substantially move a distance of approximately 253 light years. In this way, the starship at a quasi-light velocity can travel to the stars 253 light years away from us in just 100 hours.

The above numerical estimation depends on the velocity of starship. For instance, in the case of $V_S = 0.99999c$, we get

$$\Delta t_{ERS} = \Delta t_{EHS} \times 316, \quad \Delta t'_{HS} = \Delta t_{EHS} \times 1.4. \tag{11}$$

On the contrary, in the case of $V_S = 0.999......999c$, a gap between Δt_{ERS} and Δt_{EHS} rapidly increases. That depends on how the starship can be accelerated to nearly the velocity of light.

Star Flight for Stellar System

Next, a comparison is made between interstellar travel by Special Relativity and Hyper-Space Navigation. The condition is the same for both cases of navigation, that is, the distance between the earth and the star is 410 light years (i.e., Pleiades star cluster) and the velocity of starship is 0.99999c.

[Special Relativity allows the following (see Figure 6)]:

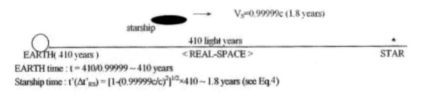

Figure 6. Interstellar Travel by Special Relativity.

A starship can travel to star 410 light years distant from us in 1.8 years. However, there exists a large problem as is well known, i.e., the twin or time paradox. If the starship travels at a velocity of 0.99999c, it will arrive at the Pleiades star cluster 1.8 years later. It will seem to the crews in the starship that only 1.8 years have elapsed. But to the people on earth it will have been 410 years. Namely, since the time gap between starship time and earth time is so large, the crew coming back to the earth will find the earth in a different period. This phenomenon is true in our Real-Space. Interstellar travel by this method is non-realistic, i.e., it would just be a one-way trip to the stars.

[Hyper-Space Navigation allows the following (see Figure 7)]:

Hyper-Space Navigation

Starship time: t'(Δt'$_{HS}$) = 1.8 years (see Eq.6)

EARTH time: t(Δt$_{EHS}$) = $(1/[1 + (0.99999c/c)^2]^{1/2}) \times 1.8 = (1/\sqrt{2}) \times 1.8 \sim 1.3$ years (see Eq.5)

Range: L = $0.99999c \times 1.3 \times ([1 + (0.99999c/c)^2]^{1/2}/[1-(0.99999c/c)^2]^{1/2})$ (see Eqs.7,9) = $0.99999c \times 1.3 \times 316 \sim 410$ light years.

Figure 7. Interstellar Travel by Hyper-Space Navigation.

A starship can travel to the stars 410 light years distant in 1.8 years. During Hyper-Space navigation of 1.8 years, just 1.3 years have passed on the earth. Therefore, the time gap between starship time and earth time is suppressed. After all, the range and travel time of starship is the same for both kinds of navigation, and travel to the stars 410 light years away can occur in just 1.8 years in both cases. However, by plunging into Hyper-Space featuring an imaginary time, i.e., a Euclidean space property, just 1.3 years, not 410 years, have passed on the earth. There is no time gap and no twin or time paradox such as in Special Relativity. Additionally, a starship can travel to the star Sirius 8.7 light years distant us in 0.039 years (14 days). During Hyper-Space navigation of 14 days, just 0.028 years (9 days) have passed on the earth.

Figure 8 shows such a realistic method for the interstellar travel using Hyper-Space navigation system (i.e., Time Hole; Figure 4). In order to reach the target star, the starship which left the earth at a velocity of approximately 0.1c to 0.2c moves and escapes completely from the Solar System (with Figure 8). After that, the starship is accelerated to nearly the speed of light in Real-Space and plunges into Hyper-Space at point A. In Hyper-Space, the time direction is changed to the imaginary time direction and the imaginary time direction is at right angles to real time. The course of starship is in the same direction, i.e., x-axis.

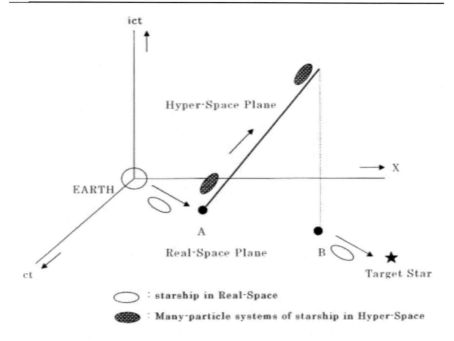

Figure 8. Interstellar travel to the star.

With the help of Eqs. (5), (7) and (9), the crew can calculate the range by the measurement of starship time. After the calculated time has just elapsed, the starship returns back to Real-Space from Hyper-Space at a point B nearby the stars. Afterward, the starship is decelerated in Real-Space and reaches the target stars. It is immediately seen that the causality principle holds. Indeed, the starship arrives at the destination ahead of ordinary navigation by passing through the tunnel of Hyper-Space (Time Hole). The ratio of tunnel passing time to earth time is 1.4:1 and both times elapse. Hyper-Space navigation method can be used at all times and everywhere in Real-Space without any restrictions to the navigation course.

This implies that Real-Space always coexists with Hyper-Space as a parallel space. The factor that isolates Real-Space from Hyper-Space consists in the usual-experience "real time" of the former as opposed to the "imaginary time" characterizing the latter. And each space is isolated by the potential barrier (Figure 5).

In general, in case that a diverse two kinds of phase space coexist or adjoin, a potential barrier shall exist to isolate these two kinds of phase space. Starship shall overcome the potential barrier by some methods. One and only difference is either real time or imaginary time. The Real-Space (3 space axes and 1 time axis) and Hyper-Space (3 space axes and 1 imaginary time axis) coexist independently and in parallel; although as if the parallel space-time exists as a five dimensional space-time (3 space axes and 2 time axes), it is not a five dimensional flat manifold. Incidentally, principle of the constancy of the speed of light, which is the strain rate of space, is established without distinguishing between real space and hyper space.

Concerning a concept on technical method of plunging into Hyper-Space and returning back to Real-Space, the following study is necessary: 1) Many-Particle Systems for Starship, 2) Wave function of Starship by Path Integrals, 3) Quantum Tunneling Effect, 4) Reduction of Wave function, 5) Starship Information Content Restoring.

While the conceptual framework discussed above is highly speculative, it is in the wake of most of the current international trends on the subject of "Interstellar Travel". Indeed, the problem of interstellar travel consists much more in a navigation theory than in propulsion theory, as there is no propulsion theory, capable of causing a starship to travel at a velocity faster than the speed of light.

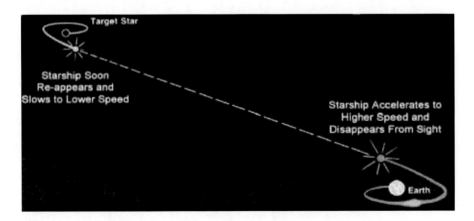

Figure 9. Interstellar Travel to the Star. (H.D. Froning).

Also, Figure 9 schematically shows the navigation of Figure 8.

Starship accelerates away from Earth, disappears and re-appears after Hyper-Space navigation. But during these Hyper-Space navigation of disappearance, the starship, in effect, leaps high above space-time and over stupendous distances to reach speeds that are substantively billions of times light-speed.

Starship flight can also be viewed from the perspective of an Earth observer who is watching a starship fly away – accelerating in the direction of its target (a planet in another solar system) and then vanishing from sight as its initial acceleration ends. The starship then re-appears after Hyper-Space navigation - at the speed it disappeared at. But the starship is now suddenly 400 light-years away - very near to its destination. By plunging into Hyper-Space featuring an imaginary time (i.e., Imaginary Time Hole), the starship detours the imaginary time tunnel (see Figure 4), apparently exceeds the speed of light.

Comparison of Wormhole and Time-Hole

Finally we compare the navigation features of the wormhole and the time-hole as shown in Table 1.

Table 1.

	Wormhole	Time-hole
method	Figure 3	Figure 4
Features of navigation	★Wormhole method is unknown where to go and how to return. ★Wormhole location unknown. ★FREE use is impossible anytime and anywhere; Limited navigation.	★ Hyper-Space navigation (Time-hole) method can be used at all times and everywhere without any restrictions to the navigation course.
Disadvantage	★ Size of wormhole is smaller than the atom, i.e., ~10⁻³⁵m and moreover the size is predicted to fluctuate theoretically due to instabilities. ★ Solution of wormhole includes a singularity, this navigation method theoretically includes fundamental problems. ★Energy necessary to expand wormhole size.	★Real-Space always coexists with Hyper-Space as a parallel space. Each space is isolated by potential barrier (Fig.5). One and only difference is either real time or imaginary time. ★Nothing in particular: TBD

Both navigation methods allow interstellar travel in a short period of time, but the features of the navigation, theoretical and technical issues are different.

Just recently (Feb. 23, 2017), NASA announced that seven planets resembling the Earth were discovered at 39 light years distant from us. Three of these planets are firmly located in the habitable zone, the area around the parent star where a rocky planet is most likely to have liquid water.

Seven Earth-sized planets have been observed by NASA's Spitzer Space Telescope around a tiny, nearby, ultra-cool dwarf star called TRAPPIST-1 (Figure 10). Three of these planets are firmly in the habitable zone.

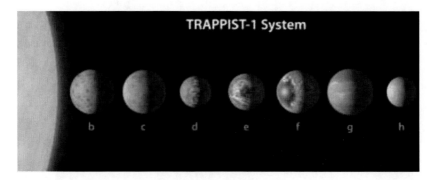

Figure 10. TRAPPIST-1 System (NASA Homepage).

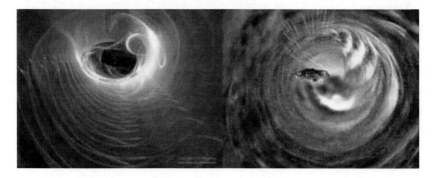

Figure 11. Interstellar travel using Hyper-Space Navigation (Time-Hole).

This exoplanet system at 39 light-years from Earth seems to be relatively close to us, but we cannot go to there due to lack of advanced propulsion system and navigation system. We need the practical space travel means which combines both space propulsion theory with space navigation theory.

As described above, a plunging into Hyper-Space characterized by imaginary time would make the interstellar travel possible in a short time. We may say that the present theoretical limitation of interstellar travel by Special Relativity is removed. The Hyper-Space navigation theory discussed above would allow a starship to start at any time and from any place for an interstellar travel to the farthest star systems, the whole mission time being within human lifetime (Figure 11).

8.4. METHODS OF HYPER-SPACE NAVIGATION

Here let us consider the concrete method of plunging into Hyper-Space and returning back to Real-Space [4, 7]. The following discussion is merely a trial calculation based on one hypothetical assumption.

Many-Particle Systems for Starship

A plunging into Hyper-Space from Real-Space can be performed everywhere in Real-Space, whenever the technical condition of starship is ready. Namely, the starship can plunge into Hyper-Space at any time without restriction of navigation course. This implies that Real-Space always coexists with Hyper-Space as a parallel space. A factor which isolates Real-Space from Hyper-Space is a usual real time and imaginary time. Such a state may be analogous to a state of de Sitter space. In general, when diverse two kinds of phase space coexist or adjoin, a potential barrier shall exist to isolate these two kinds of phase space. Therefore, the starship shall overcome the potential barrier. Next, we consider the value of above-stated potential barrier.

When the starship reaches near the velocity of light, a space as a continuum reaches at the limit of the values, i.e., a fracture point, and begins the crack. To fracture a space, the starship shall give its kinetic energy to the space as an external force. If this energy exceeds the crack energy of space, then the crack begins.

Hence, the space is to be fractured rapidly by this crack growth rate. However, this fracture is localized.

The energy (E_k) provided by starship is

$$E_K = 1/2 \cdot Mc^2, \tag{12}$$

where M is the mass of starship and c is the velocity of light.

The value of E_k varies in mass. However, since the value of potential barrier shall be constant, some standard constant value of mass is required.

Here, we adopt the Planck mass m_{PL} which is the maximum mass in elementary particle. The Planck mass is given by

$$m_{PL} = (\hbar c / G)^{1/2} = 2.2 \times 10^{-8} kg, \tag{13}$$

where $\hbar$ equals to the Planck constant divided by 2π, c is the velocity of light and G is the gravitational constant.

The Planck mass consists of only fundamental constant and plays a significant role for the unification of all interactions.

Above Planck mass gives a Planck energy E_{PL} to be

$$E_{PL} = m_{PL} c^2 = (\hbar c^5 / G)^{1/2} = 1.9 \times 10^9 J. \tag{14}$$

In addition, the Planck length L_{PL} and Planck time t_{PL} are given by respectively;

$$L_{PL} = (G\hbar / c^3)^{1/2} = 1.6 \times 10^{-35} m, \tag{15}$$

$$t_{PL} = (G\hbar/c^5)^{1/2} = 5.4 \times 10^{-44} s. \tag{16}$$

Above Planck length and Planck time are the shortest length and time which have physical meaning in quantum theory. Furthermore, these values, i.e., m_{PL}, E_{PL}, L_{PL}, t_{PL} are essential constants which play a significant role in quantum cosmology.

From above, the energy of potential barrier V_{R-H} is given by

$$V_{R-H} = 1/2 \cdot m_{PL} c^2 = 9.8 \times 10^8 J. \tag{17}$$

Here, we assumed the kinetic energy of Planck mass of light velocity as the potential energy of the barrier.

Let us suppose that the starship of mass M is formed a fine-grained structure of N- Planck masses to be

$$M = N \cdot m_{PL}. \tag{18}$$

The starship formed a fine-grained structure shall maintain the shape as many-particle systems to recreate the structure of starship existing in the initial stage. It is necessary to subdivide the starship into the size of mass that is recreated the initial structure. Because, it is impossible to recreate if the starship is subdivided into the size of atom or molecule. Therefore, the starship is composed of Planck mass of N particles. It is necessary for the starship to be formed a fine-grained structure in order to treat the starship as a many-particle systems by some technical methods. Thus, we can apply a quantum tunneling effect to the starship.

From above, the kinetic energy of starship is represented by

$$E_K = 1/2 \cdot Mc^2 = 1/2 \cdot (m_{PL} \cdot N)c^2 = (1/2 \cdot m_{PL} c^2) \cdot N. \tag{19}$$

Accordingly, the starship of mass M can be transformed to Planck mass of N.

From above discussion, the potential barrier shall be constant for all the massive body, which can be obtained by Eq. (17).

Penetration of Potential Barrier by Quantum Tunneling

Let us suppose that the thickness of potential barrier is a Planck length. This assumption is something like the potential thickness of de Sitter cosmological model. The Planck length is considered as a fundamental constant of space-time. By the method of a fine-grained structure technology, the mass of starship is subdivided into the Planck mass of N. Even if the energy of each particle is less than potential barrier, the particle can tunnel through the barrier by the quantum tunneling effect. By quantum tunneling effect, the starship as a many-particle systems can plunge into Hyper-Space without fracture of space, even if its velocity is less than the velocity of light ($V_S < C$). In the case of the energy of particle less than potential barrier ($E < V$), the transmissivity T is given by

$$T = [1+V^2 \sinh^2 \alpha d / 4E(V-E)]^{-1}, \quad \alpha = [2m(V-E)]^{1/2}/\hbar, \qquad (20)$$

where $m(kg)$ is the mass of particle, $V(J)$ is the height of potential barrier, $E(J)$ is the energy of particle and d(m) is the thickness of potential barrier.
If $\alpha d < 1$, the following approximate equation is obtained

$$T = [1+V^2 m d^2 / 2E\hbar^2]^{-1}. \qquad (21)$$

If the shape of potential barrier is not square potential but like Gaussian shape, we get the following by WKB approximation;

$$T = \exp[-2\int_a^b \{2m(V(x)-E)\}^{1/2}/\hbar \cdot dx] \approx \exp[-2\alpha d]. \qquad (22)$$

With the help of Eqs. (20), (21), and (22), let us estimate the transmissivity.

Substituting the following values for Eqs. (20) and (21), then

$V = 1/2 \cdot m_{PL} c^2 = 9.90000000 \times 10^8 J, d = L_{PL} = 1.6 \times 10^{-35} m, V_S = 0.999999999 c$

$E = 1/2 \cdot m_{PL} V_S^2 = 9.89999998 \times 10^8 J, V - E \approx 1.98 J,$

and we get $\alpha d = 4.5 \times 10^{-5} < 1$, and hence we obtain **T = 0.8**. Here we assumed the velocity of Vs is 0.999999999c.

If we use Eq. (22), the result is

$T = e^{-2\alpha d} = 0.99991 \approx 1.$

In the case of WKB approximation, which the potential energy changes slowly as a function of position, the starship composed of many-particle systems can tunnel through this potential barrier and plunge into Hyper-Space.

And also, in the case of square potential, almost particles can tunnel through this potential barrier (T = 0.8). By doing repetition, all particles can tunnel through this potential barrier.

As can be seen from above, the transmissivity depends on the mass. If the mass $m = 10 m_{PL}$, $100 m_{PL}$, $1/10 \cdot m_{PL}$, $1/100 \cdot m_{PL}$, the transmissivity T becomes $T = 0.28$, $T = 0.038$, $T = 0.975$ and $T = 0.9975$ respectively, using Eq. (20) or (21).

Therefore, the large mass cannot tunnel through the potential barrier. Anyway, the quantum tunneling effect has the advantage of plunging into Hyper-Space without the fracture of space attaining near the velocity of light.

Navigation Scenario between Real-Space and Hyper-Space

Figure.12 shows the navigation scenario of starship passing through Hyper-Space region. Although the starship is a massive body of M at a certain time, the starship is formed a fine-grained structure as a many-particle systems of $m_{PL} \times N$. The wave function of starship is required at the time. To do so, the starship turns on a fine-grained structure technology.

Hyper-Space Navigation

After that, the starship composed of many-particle systems plunges into Hyper-Space.

Then, the starship turns off a fine-grained structure technology and continues the travel in Hyper-Space. In order to jump out from Hyper-Space, the starship turns on a fine-grained structure technology again and plunges into Real-Space by quantum tunneling effect. After that, the starship turns off a fine-grained structure technology again, then decelerates and continues the travel in Real-Space.

According to the quantum mechanics, a passage through a narrow region makes the future position uncertain. From the uncertainty principle, we have

$$\Delta P \cdot \Delta x = \hbar, \tag{23}$$

where ΔP is the uncertainty in the momentum, and Δx is the uncertainty in the region.

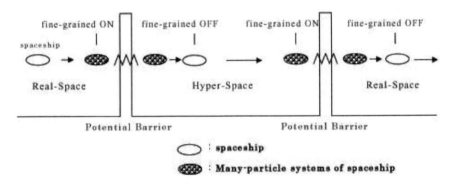

Figure 12. Navigation scenario of Starship.

From Eq. (23), we get

$$\Delta V = \hbar /(m \cdot \Delta x), \tag{24}$$

where ΔV is the velocity uncertainty, and m is the mass of particle.

We can say that a passage through a narrow region Δx causes a velocity uncertainty ΔV whose size is Eq. (24). Let us now apply the Planck unit to above equation.

Substituting the Planck mass and Planck length as thickness of barrier for Eq. (24), we get

$$\Delta V = \hbar/(m_{PL} \cdot L_{PL}) = \hbar/[(\hbar c/G)^{1/2} \cdot (G\hbar/c^3)^{1/2}] = \hbar/(\hbar/c) = c. \quad (25)$$

Eq. (25) indicates that even if the velocity of many-particle systems, i.e., starship is less than the velocity of light, the starship may achieve the velocity of light c by passing through the potential barrier. Therefore, the starship may keep the velocity of light c in Hyper-Space. A range of velocity V_S is c (velocity of light).

Hence, we may also consider that even if the starship is at rest, by turning on a fine-grained structure technology, the starship can plunge into Hyper-Space with the velocity of light. Therefore, it may not be necessary to accelerate the starship to velocity near that of light.

The time of passing through the potential barrier, from Eq. (15), we get

$$t = L_{PL}/c = (G\hbar/c^3)^{1/2}/c = (G\hbar/c^5)^{1/2} = t_{PL}. \quad (26)$$

Namely, the passing time is Planck time itself.

In addition, we have another form of uncertainty principle, that is

$$\Delta E \cdot \Delta t = \hbar, \quad (27)$$

where ΔE is the uncertainty in the energy, and Δt is the uncertainty in the time.

Substituting of Planck time into Eq. (27) gives $\Delta E = 1.9 \times 10^9$ J. This value is Planck energy E_{PL} (see Eq. (14)).

If above huge energy can be derived from passing through the potential barrier, we can avoid the following difficult problem:

1) The mass of any object would become infinite ($m \to \infty$) at near the velocity of light and the structure of starship or crew would be broken.
2) How can we get the power source of vast energy to accelerate any object to velocity near that of light?

Finally, let us supplement the properties of Hyper-Space with a few more words on referring to Figure 13. The Real-Space offered by Minkowski metric and Hyper-Space offered by Euclidean metric coexist, that is, the parallel space exists. And each space is isolated by potential barrier. The fracture of continuity of space means the crush of this potential barrier.

Hyper-Space shall be also continuum like Real-Space. One and only difference is either real time or imaginary time.

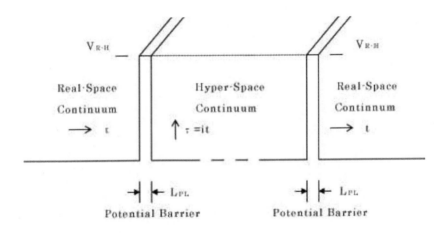

Figure 13. Properties of Hyper-Space.

Wave Function of Starship by Path Integrals

The quantum tunneling is the quantum effect that the matter passes through the inaccessible region by its wave function. Forming a fine-grained structure as many-particle systems implies the matter wave. In quantum mechanics, since giving its wave function specifies the state of system, we consider here the wave function of starship by using the path integral approach. On referring to Figure 14, $\varphi(x_{a1}, \cdots x_{aN}, t_a)$ is the wave function of many-particle systems of N particles when the starship is formed a fine-grained structure at the point a. The wave function $\varphi(x_{b1}, \cdots x_{bN}, it_b)$ of many-particle systems after passing through the potential barrier is given by using the path integral expression;

$$\varphi(x_{b1}, \cdots x_{bN}, it_b) = \int_{-\infty}^{+\infty} [dx_{aN}] K(x_{b1}, \cdots x_{bN}, it_b; x_{a1}, \cdots x_{aN}, t_a) \varphi(x_{a1}, \cdots x_{aN}, t_a), \quad (28)$$

where $\int [dx_{aN}] = \int \cdots \int dx_{a1} dx_{a2} \cdots dx_{aN}$.

Let d and c represent the position of potential barrier, and let t_d and it_c (imaginary time) be the time of position of d and c. The total amplitude which goes from the point in space-time (x_a, t_a) to (x_b, it_b), i.e., Feynman Kernel $K(b,a)$ is given by

$$K(b,a) = \int\int [dx_{cN}][dx_{dN}] K(b,c) K(c,d) K(d,a). \quad (29)$$

All paths between Real-Space and Hyper-Space are divided into two parts.

The time is real time for between a and d, and imaginary time for between c and b. Finally, as point d comes closer and closer to point c, the real time t gets closer and closer to imaginary time it, i.e., analytic continuation.

Hyper-Space Navigation 219

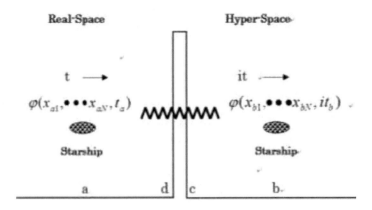

Figure 14. Wave function of Starship.

Each kernel is represented as follows:

$$K(d,a) = K(x_{d1},\cdots x_{dN},t_d;x_{a1},\cdots x_{aN},t_a) = \int_a^d dx \cdot \exp[i/\hbar \cdot \int_{ta}^{td} dt L(\dot{x},x,t)],$$

$$K(b,c) = K(x_{b1},\cdots x_{bN},it_b;x_{c1},\cdots x_{cN},it_c) = \int_c^b dx \cdot \exp[-1/\hbar \cdot \int_{tc}^{tb} dt L(\dot{x},x,it)], \quad (30)$$

where L is the Lagrangian for system.

In the case of starship, the Lagrangian is a free particle system and given by

$$L = \sum 1/2 \cdot m_{PL} \dot{x}_N^2. \quad (31)$$

Taking the limit as (d-c) approaches zero, we get

$$K(b,a) = \int [dx_{d\to c,N}] K(b,c) K(d \to c,a),$$

$$\varphi(x_{b1},\cdots x_{bN},it_b) = \int_{-\infty}^{+\infty} [dx_{aN}] K(b,a) \varphi(x_{a1},\cdots x_{aN},t_a). \quad (32)$$

The wave function of starship is to be found out as above. To find out the kernel is equal to solve the following Schrödinger equation

$$H\varphi(r_1, \cdots r_N) = i\hbar \frac{\partial \varphi(r_1, \cdots r_N)}{\partial t}, \qquad (33)$$

where H is the Hamiltonian operator.

The wave function of starship in Hyper-Space can be represented by the wave function in Real-Space.

REFERENCES

[1] Forward, R.L., "Space Warps: A Review of One Form of Propulsionless Transport", *JBIS*, **42**, pp.533-542 (1989).

[2] Froning Jr, H.D., "Requirements for Rapid Transport to the Further Stars", *JBIS*, **36**, pp.227-230 (1983).

[3] Froning Jr, H.D., *The Halcyon Years of Air and Space Flight: And the Continuing Quest*, published in May. 13, 2016 (LAMBERT Academic Publishing).

[4] Minami, Y., "Hyper-Space Navigation Hypothesis for Interstellar Exploration (IAA.4.1-93-712)", *44th Congress of the International Astronautical Federation (IAF)*, 1993.

[5] Minami, Y., "Travelling to the Stars: Possibilities Given by a Spacetime Featuring Imaginary Time", *JBIS*, **56**, pp.205-211 (2003).

[6] Minami, Y., "A Perspective of Practical Interstellar Exploration: Using Field Propulsion and Hyper-Space Navigation Theory", in the proceedings of *Space Technology and Applications International Forum (STAIF-2005)*, edited by M. S. El-Genk, AIP Conference Proceedings 746, Melville, New York, 2005, pp. 1419-1429.

[7] Minami, Y., *A Journey to the Stars – By Means of Space Drive Propulsion and Time-Hole Navigation –*, published in Sept. 1, 2014 (Lambert Academic Publishing).

[8] Minami, Y., "Interstellar travel through the Imaginary Time Hole", *Journal of Space Exploration* 3, 2014: 206-212.

[9] Minami, Y., "Space propulsion physics toward galaxy exploration", *J Aeronaut Aerospace Eng* 4: 2; 2015.

[10] Hawking, S., *A Brief History of Times*, Bantam Publishing Company, New York, 1988.

[11] Hawking, S., *Hawking on Bigbang and Black Holes*, World Scientific, 1993.

[12] Williams C, Cardoso JG, Whitney CK, Minami Y, Mabkhout SA, et al., *Advances in general relativity research*, Nova Science Publishers, 2015.

CONCLUSION

By making obvious and generally accepted assumptions based on existing established physics, we have been able to introduce the possibility of intergalactic exploration using field propulsion.

It is concluded that spaceflight propulsion must ultimately almost entirely consist of actions and reactions of fields, not combustion and expulsion of masses - with a vehicle's "fuel" or "propellant" being mainly a cryogenic "working fluids" for removal of waste heat.

Though this book mainly describes the development of propulsive force or thrusting acceleration by actions of fields, these actions require energies developed by on-board field power systems. For the only propulsive fields we can currently create are electro-magnetic ones, which require energy from electric and/or magnetic power generating systems. So, the most promising field propulsion systems are those whose electro-magnetic emissions could interact favorably with the vehicle flight medium (air atoms or molecules or gravitational metric of space time or zero-point quantum energies of space) to develop acceleration or thrust. And those emissions that can also interact with this medium to harvest its energies for the field propulsion system use.

Evolution of field propulsion will probably be gradual – with field propulsion possibly augmenting (rather than totally replacing) matter-consuming propulsion in its initial use. But if humanity has a real destiny in space, so does nearly propellant-less field propulsion. For only it can reduce spaceflight fuel and cost and flight hazard by factors of 5 to 10 – factors needed for giving the precious, inspiring experience of spaceflight to thousands of Earth's people by the middle of this century and of achieving cost-effective interplanetary flight to other worlds – and, by this century's end, - conceivably achieve sufficient field propulsion perfection to embark on interstellar flight to a distant star.

Then, slow, but reasonably economical, reliable and safe field-propelled ships that could reach solar-system escape velocity, could slowly colonize our entire Milky-Way galaxy if nanotechnology and molecular manufacturing would allow their self-replicating into an ever expanding fleet that could colonize most of our own galaxy in 100 to 200,000 years. Field-propelled ships that could reach almost light-speed c could conceivably colonize our galaxy in less than 100 years. And, if swiftly, accelerating faster-than-light field-propelled ships with hyperspace navigation ever come to pass, our entire galaxy could be colonized and connected with other galactic civilizations by faster-than-light interstellar commerce with them.

Right now this rosy picture of cosmic exploration by field-powered and propelled ships would seem a completely impossible dream. But when the world finally realizes its very survival depends on stopping global warming by stopping almost all combusting and emission of matter; and when it realizes this can be done only by energy and transport by nearly propellant-less field-power and propulsion, there will hopefully be a world-uniting international initiative to perfect field power and propulsion during this century's last critical years – a World-uniting initiative to deliver our planet from peril and begin the first human steps to the stars - by propellant-less power and field propulsion.

APPENDICES

APPENDIX A: OUTLINE OF GENERAL RELATIVITY

In this section, we run through the nutshell of General Relativity to understand the space drive propulsion theory in brief.

General Relativity is the geometric theory of gravitation formulated by Albert Einstein in 1916, which was an extension of Special Relativity. It unifies special relativity and Newton's law of universal gravitation, and describes gravity as a geometric property of space and time, or space-time. In particular, the curvature of space-time is directly related to the four-momentum (mass-energy and linear momentum) of whatever matter and radiation are present. The relation is specified by the Einstein field equations, a system of partial differential equations. Many predictions of general relativity differ significantly from those of classical physics, especially concerning the passage of time, the geometry of space, the motion of bodies in free fall, and the propagation of light. Einstein's theory has important astrophysical implications, that is, general relativity is the basis of current cosmological models of a consistently expanding universe.

What is gravity? The answer is that the gravity is explained by curved space-time. From the standpoint of mathematical method, Riemannian geometry and tensor analysis are required. Especially, the metric tensor is important and fundamental. The notion of tensor is extended from the notion of vector. The brief explanation is described below.

The quantities $T_{iklm...}{}^{rst...}$ in which the indices can, independently, take on the values 1,2,3,4 are called tensor components. They are, in particular, called covariant components for the indices $iklm...$ and contravariant for the indices $rst...$.

The number of indices which the components have is called the rank of the tensor. Tensors of first rank are also called vectors. The example of vector is the coordinates x^i of a point (x^1, x^2, x^3, x^4 for 4 dimensional spaces). A vector is a tensor whose components are equal in number to the dimensionality of the space in which it is defined. Tensors are mathematical objects that represent a generalization of the vector concept. As well, a tensor of zero rank is called scalar.

A tensor of second rank a_{ik} and a vector x^k can be combined to give the vector $z_i = a_{ik} x^k$. There are many different types of tensors, such as the strain tensor describes the local distortion of material, metric tensor, and curvature tensor, which has twenty components, describes the deviation of space-time continuum from flatness.

Since the components of vector is three in three dimensional space (F_i: $i = 1,2,3$), the components of tensor of second rank is enlarged to nine, as (F_{ij}: $i = 1,2,3$, $j = 1,2,3$). F_{ij} is called second rank tensor. As might be expected, in the case of tensor, there are third rank tensor, ... n-th rank tensor. The metric tensor is usually expressed g_{ij}.

The metric tensor of Special Relativity which shows flat space-time is a constant, and takes constant value regardless of a place. However, metric tensor of the curved space is a function of the place which changes a value by a place.

In General Relativity, mass particles move the geodesic line of space-time. A geodesic line is a curve which connects two on a curved surface with the shortest distance. Although the trajectory of a particle serves as a straight line in flat space, the trajectory of a particle serves as a curve in the curved space-time with gravity. That is, a particle moves in accordance with the following geodesic equation:

$$\frac{d^2 x^\mu}{d\tau^2} + \Gamma^\mu{}_{\nu\lambda} \frac{dx^\nu}{d\tau} \frac{dx^\lambda}{d\tau} = 0 \tag{A.1}$$

In the curved space, a result from which the trajectory differed is brought by which course even if particles move a closed curve, is chosen. Even if it goes around, it cannot return to the starting point of a basis.

For example, in the space-time at which it turned in on the earth, even if it carries out parallel displacement of the vector on between space-time, the results of parallel displacement differ by along which route it passed. In order to explain this phenomenon, curvature tensor (Riemann tensor) plays a role.

A coordinate system serves as a curvilinear coordinate system fundamentally. Thus, it is the greatest feature of the General Relativity to have the curvature. The Riemannian geometry which shows the geometrical structure of such curved space-time mathematically serves as an important mathematical means. The physical space in which we live in is the Riemann space-time.

Then, what this curvature does to a reason? If energies, such as mass energy of a mass object, electromagnetic energy, and thermal energy, exist, space-time will be curved. It is called a gravitational field equation alias Einstein equation to express this relation.

$$R^{ij} - \frac{1}{2} \cdot g^{ij} R = -\frac{8\pi G}{c^4} \cdot T^{ij} \tag{A.2}$$

where R^{ij} is the Ricci tensor, R is the scalar curvature, g^{ij} is metric tensor, G is the gravitational constant, c is the speed of light, and T^{ij} is the energy momentum tensor.

The left side shows the bend condition of space-time and the right-hand side shows the energy source to generate the curvature of space-time.

The basic preconditions of the General Relativity are the general principle of relativity and an equivalence principle. Firstly, all the organic law of physics is equally materialized in all the coordinate systems, i.e., the general principle of relativity is that a physical law is invariance to general coordinate conversion. Secondary, an equivalence principle is saying that the size of gravity is changeable with suitable coordinate conversion, and the true gravity by the earth and the apparent force of the force of inertia by coordinate conversion cannot be identified.

The general laws of physics can be expressed in a form which is independent of the choice of space-time coordinates. For this reason, the tensor analysis is effective.

As described above, the flat space-time of Minkowski for Special Relativity is to be replaced by a curved space-time, and the curvature is to be responsible for gravitational effect in General Relativity.

The difference can be explained as whether space is curved or not, that is, whether 20 independent components of Riemann curvature tensor are zero or not. After all, the existence of space curvature determines whether the object drops straight down or not. Although the space curvature at the surface of the Earth is very small, i.e., $3.42 \times 10^{-23} (1/m^2)$, it is enough value to produce 1G ($9.8 m/s^2$) acceleration. On the contrary, the space curvature in the universe (flat space) is zero; therefore any acceleration is not produced. Accordingly, if the space curvature of a localized area including object is controlled to $3.42 \times 10^{-23} (1/m^2)$ curvature and curved space region, the object moves receiving 1G acceleration in the universe. Because the intensity of acceleration produced in curved space is proportional to both spatial curvature and the size of curved space.

The square of the infinitesimal distance "ds" between two infinitely proximate points x^i and $x^i + dx^i$ is given by equation of the form

$$ds^2 = g_{ij} dx^i dx^j \qquad (A.3)$$

where g_{ij} is metric tensor.

The metric tensor g_{ij} is a quantity defined geometrical character of space and is functions of x^i in general. The metric tensor is constant in the case of flat space (i.e., Special Relativity represented by the Minkowski metric η_{ij}). In a curved Riemannian space, the property of space depends on the metric

Appendices

tensor g_{ij}. The infinitesimal distance ds^2 is given by metric tensor g_{ij}, and also metric tensor g_{ij} determines Riemann connection coefficient Γ^i_{jk}, furthermore Riemann curvature tensor $R^p_{ijk} = (R_{pijk})$.

Thus, the geometry of space is determined by metric tensor g_{ij}. Riemann curvature tensor is represented as follows:

$$R_{\mu\nu kl} = \frac{1}{2} \cdot (g_{\mu l,\nu k} - g_{\nu l,\mu k} - g_{\mu k,\nu l} + g_{\nu k,\mu l}) + \Gamma_{\beta\mu l}\Gamma^\beta_{\mu k} - \Gamma_{\beta\mu k}\Gamma^\beta_{\nu l} \quad (A.4)$$

$$\Gamma^r_{ij} = \frac{1}{2} \cdot g^{rk}(g_{jk,i} + g_{ki,j} - g_{ij,k}) \quad (A.5)$$

Here, we use the notation $g_{jk,i}$ for $\dfrac{\partial g_{jk}}{\partial x^i}$.

As described above, Riemann curvature tensor $R_{\mu\nu kl}$ consists of fundamental metric tensor $g_{\mu\nu}$, therefore the structure of space-time is determined by metric tensor $g_{\mu\nu}$.

The solution of metric tensor $g_{\mu\nu}$ is found by gravitational field equation described above as the following:

$$R^{ij} - \frac{1}{2} \cdot g^{ij} R = -\frac{8\pi G}{c^4} \cdot T^{ij} \quad (A.6)$$

Furthermore, we have the following relation for scalar curvature R:

$$R = R^\alpha{}_\alpha = g^{\alpha\beta} R_{\alpha\beta}, \quad R^{\mu\nu} = g^{\mu\alpha} g^{\nu\beta} R_{\alpha\beta}, \quad R_{\alpha\beta} = R^j{}_{\alpha j\beta} = g^{ij} R_{i\alpha j\beta} \quad (A.7)$$

Ricci tensor is represented by

$$R_{\mu\nu} = \Gamma^{\alpha}_{\mu\alpha,\nu} - \Gamma^{\alpha}_{\mu\nu,\alpha} - \Gamma^{\alpha}_{\mu\nu}\Gamma^{\beta}_{\alpha\beta} + \Gamma^{\alpha}_{\mu\beta}\Gamma^{\beta}_{\nu\alpha} \quad (= R_{\nu\mu}) \tag{A.8}$$

If the curvature of space is very small, the term of higher order than the second can be neglected and Ricci tensor becomes

$$R_{\nu\nu} = \Gamma^{\alpha}_{\mu\alpha,\nu} - \Gamma^{\alpha}_{\mu\nu,\alpha} \tag{A.9}$$

From Eq. (A.7), the major curvature of Ricci tensor ($\mu = \nu = 0$) is calculated as follows:

$$R^{00} = g^{00}g^{00}R_{00} = -1 \times -1 \times R_{00} = R_{00} \tag{A.10}$$

As is well known, Riemannian geometry is geometry that deals with curved Riemann space, therefore, Riemann curvature tensor is the principal quantity. All components of Riemann curvature tensor are zero for flat space and non-zero for curved space. If an only non-zero component of Riemann curvature tensor exists, the space is not flat space but curved space.

Further, in a curved space, it is well known that the result of the parallel displacement of vector depends on the choice of the path and also the components of vector differ from the initial value after the vector parallel displacement is performed along a closed curve until it returns to the starting point.

General Relativity Solutions for Field Propulsion

The acceleration (α) of curved space and its Riemannian connection coefficient (Γ^3_{00}) are given by

$$\alpha = \sqrt{-g_{00}} c^2 \Gamma^3_{00}, \quad \Gamma^3_{00} = \frac{-g_{00,3}}{2g_{33}}, \tag{A.11}$$

where c: velocity of light, g_{00} and g_{33}: component of metric tensor, $g_{00,3}$: $\partial g_{00}/\partial x^3 = \partial g_{00}/\partial r$.

We choose the spherical coordinates "$ct = x^0, r = x^3, \theta = x^1, \varphi = x^2$" in space-time. The acceleration α is represented by the equation both in the differential and in the integral form. Practically, since the metric is usually given, the differential form has been found to be advantageous.

Table A1 is effective for the calculation of General Relativity.

Next, we expand on these categories as they relate to other solutions of gravitational field equation, that is, the concrete acceleration α is derived from Eq. (A.11).

Table A1. Equations effective for the calculation of General Relativity

$$\Gamma_{m,n,n} = (\Gamma_{nm,n}) = -\frac{1}{2} g_{mm,n}$$

$$\Gamma_{m,n,m} = \frac{1}{2} g_{mm,n}$$

$$\Gamma_{m,n,n} = (\Gamma_{nm,n}) = \frac{1}{2} g_{nn,m}$$

$$\Gamma_{n,n,m} = (\Gamma_{mn,n}) = \frac{1}{2} g_{mm,n}$$

$$\text{other } \Gamma_{\mu\nu\lambda} = 0 \;\; (\because g_{\mu\nu} = 0 \;\; \mu \neq \nu)$$

$$\Gamma^{m}_{mn} = \Gamma^{m}_{nm} = \frac{g_{mm,n}}{2 g_{mm}}$$

$$\Gamma^{m}_{mm} = \frac{g_{mm,m}}{2 g_{mm}}$$

$$\Gamma^{m}_{nn} = -\frac{g_{nn,m}}{2 g_{mm}}$$

$$\text{other } \Gamma^{\mu}_{\nu\lambda} = 0$$

$$R_{\mu\nu k l} = \frac{1}{2}(g_{\mu l,\nu k} - g_{\nu l,\mu k} - g_{\mu k,\nu l} + g_{\nu k,\mu l}) + \Gamma_{\beta\nu l}\Gamma^{\beta}_{\mu k} - \Gamma_{\beta\nu k}\Gamma^{\beta}_{\mu l}.$$

External Schwarzschild Solution

The metrics are given by:

$$g_{00} = -(1 - r_g/r), g_{11} = g_{22} = 1, g_{33} = 1/(1 - r_g/r), \quad \text{(A.12)}$$
$$\text{and other } g_{ij} = 0.$$

where r_g is the gravitational radius (i.e., $r_g = 2GM/c^2$).

Combining Eq. (A.11) with Eq. (A12) yields:

$$\alpha = G \cdot \frac{M}{r^2}, \ (r_g \langle \ r) \ , \tag{A.13}$$

where G is a gravitational constant and M is a total mass.

Reissner-Nordstrom Charged Mass Solution

The metrics outside of charged and spherically symmetric mass are given by:

$$g_{00} = -(1 - r_g/r + Q^2/r^2), g_{11} = g_{22} = 1, g_{33} = 1/(1 - r_g/r + Q^2/r^2),$$
and other $g_{ij} = 0$.
(A.14)

where $Q^2 = Gq^2/c^4$ ($q = electric\ charge$), $r_g = 2GM/c^2$.

Eq. (A.14) reduces to the Schwarzschild solution if electric charge "q" is zero.

Combining Eq. (A.11) with Eq. (A.14) yields:

$$\alpha = G \cdot \frac{M}{r^2} - \frac{Gq^2}{c^2 r^3} < G \cdot \frac{M}{r^2}, \ (r_g < r, Q^2 < r^2). \tag{A.15}$$

Eq. (A.15) indicates that the electric charge weakens the gravitational acceleration.

Kerr Rotating Mass Solution

The metrics outside of spinning mass are given by:

$$g_{00} = -\left(1 - \frac{r_g r}{r^2 + h^2 \cos^2\theta}\right), \quad g_{33} = \frac{r^2 + h^2 \cos^2\theta}{r^2 - r_g r + h^2}, \quad (A.16)$$

where $h = J/Mc$ (J = *angular momentum*), $r_g = 2GM/c^2$.

Eq. (A.16) reduces to the Schwarzschild solution if the angular momentum "*J*" is zero.

Combining Eq. (A.11) with Eq. (A.16) yields:

$$\alpha = G \cdot \frac{M}{r^2} \cdot \frac{(1 - h^2 \cos^2\theta/r^2)}{(1 + h^2 \cos^2\theta/r^2)^3} < G \cdot \frac{M}{r^2}, \quad (r_g < r, h^2 < r^2). \quad (A.17)$$

Eq. (A.17) indicates that the rotation weakens the gravitational acceleration.

Internal Schwarzschild Solution

The space-time metrics inside of a static, constant energy density, perfect fluid sphere are given by:

$$g_{00} = -\left[\frac{3}{2} \cdot (1 - K\rho a^2/3)^{\frac{1}{2}} - \frac{1}{2} \cdot (1 - K\rho r^2/3)^{\frac{1}{2}}\right]^2, \quad g_{33} = \frac{1}{1 - K\rho r^2/3}, \quad (A.18)$$

$g_{11} = g_{22} = 1$, *and other* $g_{ij} = 0$.

where $K = 8\pi G/c^4$, ρ is the energy density (J/m³), "*a*" is the radius of energy density (i.e., fluid boundary at $r = a$). This solution corresponds to the so-called Poisson equation. While, External Schwarzschild Solution corresponds to the so-called Laplace equation.

Combining Eq. (A.11) with Eq. (A.18) yields:

$$\alpha = \frac{GM}{a^3} \cdot r \quad (r_g < a). \tag{A.19}$$

Eq. (A.19) reduces to the Eq. (A.13), if $r = a$, and the continuity at "$r = a$" links the internal solution to external solution.

De Sitter Solution

In the latest cosmology, the terms vacuum energy and cosmological term "Λg^{ij}" are used synonymously. Λ is a constant known as the cosmological constant. The cosmological term is identical to the stress-energy associated with the vacuum energy.

Now, concerning the de Sitter cosmological model with non-zero vacuum energy (i.e., cosmological constant), the de Sitter line element is written as

$$ds^2 = -(1 - \frac{1}{3}\Lambda r^2)c^2 dt^2 + \frac{1}{1 - \frac{1}{3}\Lambda r^2} dr^2 + r^2(d\theta^2 + \sin^2\theta d\varphi^2)$$

$$\tag{A.20}$$

The metrics are given by

$$g_{00} = -(1 - 1/3 \cdot \Lambda r^2), \; g_{11} = g_{22} = 1, \; g_{33} = 1/(1 - 1/3 \cdot \Lambda r^2),$$
$$\text{and other } g_{ij} = 0 \tag{A.21}$$

The acceleration α of de Sitter solution can be obtained by combining Eq. (A.11) with Eq. (A.21)

$$\alpha = \frac{1}{3}c^2 \Lambda r \quad (1 > 1/3 \cdot \Lambda r^2). \tag{A.22}$$

The acceleration induced by cosmological constant is proportional to the distance "r" from the generative source.

Example:
The gravitational field equation including the cosmological constant Λ is given by the following equation:

$$R^{ij} - \frac{1}{2} \cdot g^{ij} R = -\frac{8\pi G}{c^4} T^{ij} + \Lambda g^{ij} .$$

Here, if we multiply both sides of Equation by g_{ij},

$$R^{ij} g_{ij} - \frac{1}{2} g^{ij} g_{ij} R = -\frac{8\pi G}{c^4} T^{ij} g_{ij} + \Lambda g^{ij} g_{ij} \rightarrow$$

$$\rightarrow R_i^i - \frac{1}{2} \times 4R = -\frac{8\pi G}{c^4} T_i^i + \Lambda \times 4 .$$

Using, $g^{ij} g_{ij} = g_i^i = \delta_i^i = 4$, $R_i^i = R$, $T_i^i = T$, then

$R - 2R = -\frac{8\pi G}{c^4} T + 4\Lambda$. Accordingly, we obtain $\frac{8\pi G}{c^4} T = R + 4\Lambda$.

In empty space with all of the components of the energy momentum tensor are equal to zero, that is, $T^{ij} = 0$ and $T = 0$, we get the following respectively

$T^{ij} = 0$, $T = 0$, then, from $0 = R + 4\Lambda$, then $R = -4\Lambda$.

Also, from $R^{ij} - \frac{1}{2} g^{ij} \times -4\Lambda = 0 + \Lambda g^{ij}$, then $R^{ij} + 2g^{ij}\Lambda = g^{ij}\Lambda$.

Namely, $R^{ij} = -g^{ij}\Lambda = -\Lambda g^{ij}$.

Hence we get the following, $R = -4\Lambda$, $R^{ij} = -\Lambda g^{ij}$.

APPENDIX B: DERIVATION OF EQ. (23) AND DERIVATION OF EQ. (30) IN CHAPTER 4

< Derivation of Eq. (23) >

Let us consider two adjacent spatial points A and B in the undeformed space, Figure 3 (chapter 4), which are the end points of a line element vector *ds*. During the deformation, point A undergoes the displacement u and moves point A', while point B experiences a slightly different displacement $u + du$ when moving to point B'.

From Figure 3 in chapter 4, we read the simple vector equation

$$ds' = ds + du = ds + g^i u_{i:j} dx^j = g_i dx^i + g^i u_{i:j} dx^j.$$ (B1)

As explained in chapter 4, the infinitesimal distance between the two near points is given by

$$ds^2 = g_{ij} dx^i dx^j,$$ (B2)

$$ds'^2 = g'_{ij} dx^i dx^j.$$ (B3)

We may now write the square of the deformed line element. Since all indices are dummies, they have been chosen so that the final result looks best. When we multiply the two factors term by term and switch the notation for some dummy pairs, we obtain:

$$ds' \cdot ds' = (g_i dx^i + g^k u_{k:i} dx^i) \cdot (g_j dx^j + g^l u_{l:j} dx^j) =$$
$$= (g_{ij} + 2 g_i \cdot g^l u_{l:j} + g^{kl} u_{k:i} u_{l:j}) dx^i dx^j.$$ (B4)

Using Eq. (B3) and Eq. (B2), from Eq. (B4), we get:

$$ds'^2 - g_{ij}dx^i dx^j = (2g_i \cdot g^l u_{l:j} + g^{kl} u_{k:i} u_{l:j}) dx^i dx^j. \tag{B5}$$

Left side of Eq. (B5):

$$g'_{ij} dx^i dx^j - g_{ij} dx^i dx^j = (g'_{ij} - g_{ij}) dx^i dx^j = r_{ij} dx^i dx^j. \tag{B6}$$

Right side of Eq. (B5):
Considering $2g_i g^l u_{l:j} = 2\delta_i^l u_{l:j} = 2u_{i:j}$ and $g^{kl} u_{k:i} u_{l:j} = u^l{}_{:i} u_{l:j} = u^k{}_{:i} u_{k:j}$, (changes of the dummy indices l→k), then,

$$(2g_i \cdot g^l u_{l:j} + g^{kl} u_{k:i} u_{l:j}) dx^i dx^j = (2u_{i:j} + u^k{}_{:i} u_{k:j}). \tag{B7}$$

Since $r_{ij} = r_{ji}$, considering $2u_{i:j} = u_{i:j} + u_{j:i}$,

$$(g'_{ij} - g_{ij}) dx^i dx^j = r_{ij} dx^i dx^j = (u_{i:j} + u_{j:i} + u^k{}_{:i} u_{k:j}) dx^i dx^j. \tag{B8}$$

Finally we obtain:

$$ds'^2 - ds^2 = r_{ij} dx^i dx^j = (u_{i:j} + u_{j:i} + u^k{}_{:i} u_{k:j}) dx^i dx^j. \tag{B9}$$

< Derivation of Eq. (30) >

Let us suppose that covariant vector $A_i(P)$ at point P(x) is transported parallel to Q(x + d1x + d2x) via path I is $A_i(Q)_I$. On the while, covariant vector $A_i(P)$ at point P(x) is transported parallel to Q(x + d1x + d2x) via path II is $A_i(Q)_{II}$. (Figure B1).

The difference of these parallel transport of vector via two paths is indicated by:

$$A_i(Q)_I - A_i(Q)_{II} = R^\rho{}_{ik\sigma}(P)A_\rho(P)d_1x^k d_2x^\sigma, \tag{B10}$$

where $R^\rho{}_{ik\sigma}(P)$ is Riemann curvature tensor at point P.

Considering $A_i(Q)_I + (-A_i(Q)_{II})$, Eq. (B10) indicates the quantum not returned of vector $A_i(P)$ in case that parallel transport of a vector $A_i(P)$ along a closed path has been employed at each segment of the loop $P \to R\ I \to Q \to R\ II \to P$ and ultimately leads back to the point of departure, i.e., original point P(x) (Figure B2).

Mathematically, Riemann curvature tensor is the result of a difference that changed the order of the covariant derivatives as seen Eq. (B11), and its non-commutative part is represented by the Riemann curvature tensor.

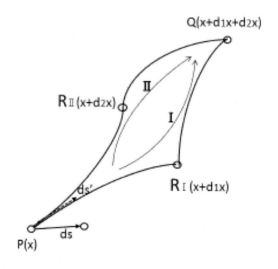

Figure B1. Parallel transport of vector via two paths.

$$X_{i:jk} - X_{i:kj} = R^p{}_{ijk} X_p. \tag{B11}$$

Let us consider point R adjacent spatial point P, Figure B2, which are the end points of a line element vector ds. If vector ds at P(x) is transported parallel to RI (x + d1x) and then to Q(x + d1x + d2x), then parallel transport from Q(x + d1x + d2x) to original point P(x) via II (x + d2x), the result is the new vector *ds'*.

Parallel transport of a vector *ds* along a closed path that ultimately leads back to the point of departure will result in a new vector *ds'* at the original point P(x); the new vector *ds'* differs from the original vector *ds*, even though the proper procedure for parallel transport has been employed at each segment of the loop. *ds'-ds* indicates the quantum not returned of vector, also is denoted by displacement vector *du*. This arises from nonzero curvature of space.

Another interpretation, two adjacent spatial points P and R in the undeformed space, Figure B2, which are the end points of a line element vector *ds*. During the deformation, *P* under goes the displacement u and moves *P'*, while *R* experiences a slightly different displacement $u + du$ when moving to *R'*.

These phenomena are equivalent, it is not possible to identify them.

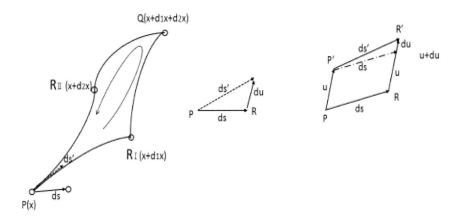

Figure B2. Parallel transport of vector along a closed path and displacement vector.

From above, we get:

$$A_i(Q)_I - A_i(Q)_{II} = ds' - ds = R^\rho{}_{ik\sigma}(P) A_\rho(P) d_1 x^k d_2 x^\sigma . \qquad (B12)$$

Since *ds'-ds = du*, infinitesimal displacement vector du is described in

$$du = du_i = u_{i;j}dx^j.\tag{B13}$$

$$A_i(Q)_I - A_i(Q)_{II} = ds' - ds = du = u_{i;j}dx^j.\tag{B14}$$

Apply a vector A_ρ (P) in Eq. (B12) to a line element vector $ds = ds_\rho$, from Eq. (B14), we get

$$u_{i;j}dx^j = R^\rho{}_{ik\sigma}ds_\rho d_1 x^k d_2 x^\sigma.\tag{B15}$$

Now let us multiply both sides of Eq. (B15) by $g^{r\rho}$, we get following:

$$g^{r\rho}u_{i;j}dx^j = R^\rho{}_{ik\sigma}ds^r d_1 x^k d_2 x^\sigma \text{, then,}$$

$$g^{r\rho}u_{i;j} = R^\rho{}_{ik\sigma}\frac{ds^r}{dx^j}d_1 x^k d_2 x^\sigma = R^\rho{}_{ik\sigma}\frac{dx^r}{dx^j}d_1 x^k d_2 x^\sigma = R^\rho{}_{ik\sigma}\delta^r_j d_1 x^k d_2 x^\sigma,$$

finally

$$g^{j\rho}u_{i;j} = R^\rho{}_{ik\sigma}d_1 x^k d_2 x^\sigma.\tag{B16}$$

Multiplying both sides of Eq. (B16) by $g_{\rho m}$,

$$g_{\rho m}g^{j\rho}u_{i;j} = g_{\rho m}R^\rho{}_{ik\sigma}d_1 x^k d_2 x^\sigma \text{, then applying, } g_{\rho m}g^{j\rho} = \delta^j_m,$$

we get:

$$u_{i;m} = R_{mik\sigma}d_1 x^k d_2 x^\sigma.\tag{B17}$$

For convenience, returns the index m to j,

$$u_{i;j} = R_{jik\sigma}d_1 x^k d_2 x^\sigma.\tag{B18}$$

Interchanging index i, j, we get:

$$u_{j:i} = R_{ijk\sigma} d_1 x^k d_2 x^\sigma. \tag{B19}$$

On the other hand, using the nature of the Riemann curvature tensor,

$$R_{jik\sigma} = -R_{ijk\sigma}. \tag{B20}$$

Subtracting Eq. (B19) from Eq. (B18), we obtain:

$$u_{i:j} - u_{j:i} = -R_{ijk\sigma} d_1 x^k d_2 x^\sigma - R_{ijk\sigma} d_1 x^k d_2 x^\sigma = -2R_{ijk\sigma} d_1 x^k d_2 x^\sigma. \tag{B21}$$

By continuum mechanics, anti-symmetric part of the displacement gradient tensor represents the rotation tensor ω_{ij},

$$\omega_{ij} = \frac{1}{2}(u_{j:i} - u_{i:j}). \tag{B22}$$

Accordingly, we obtain:

$$\omega_{ij} = \frac{1}{2}(u_{j:i} - u_{i:j}) = R_{ijk\sigma} d_1 x^k d_2 x^\sigma = R_{ijk\sigma} dA^{k\sigma}, \tag{B23}$$

where $dA^{k\sigma}$ is the area element enclosed by the vector $d_1 x^k$ and vector $d_2 x^\sigma$.

Thus, changing the dummy of incidences $i, j, k, \sigma \to \mu, \nu, k, l$, we get finally:

$$\omega_{\mu\nu} = R_{\mu\nu kl} dA^{kl}, \tag{B24}$$

where $\omega_{\mu\nu}$ is rotation tensor, dA^{kl} is infinitesimal areal element.

APPENDIX C: CURVATURE CONTROL BY MAGNETIC FIELD

Let us consider the electromagnetic energy tensor M^{ij}. In this case, the solution of metric tensor g_{ij} is found by

$$R^{ij} - \frac{1}{2} \cdot g^{ij} R = -\frac{8\pi G}{c^4} \cdot M^{ij}. \tag{C1}$$

Eq. (C1) determines the structure of space due to the electromagnetic energy.

Here, if we multiply both sides of Eq. (C.1) by g_{ij}, we obtain

$$g_{ij}\left(R^{ij} - \frac{1}{2} \cdot g^{ij} R\right) = g_{ij} R^{ij} - \frac{1}{2} \cdot g_{ij} g^{ij} R = R - \frac{1}{2} \cdot 4R = -R, \tag{C2}$$

$$g_{ij}\left(\frac{-8\pi G}{c^4} \cdot M^{ij}\right) = -\frac{8\pi G}{c^4} \cdot g_{ij} M^{ij} = \frac{-8\pi G}{c^4} \cdot M_i^i = \frac{-8\pi G}{c^4} M. \tag{C3}$$

The following equation is derived from Eqs. (C2) and (C3)

$$R = \frac{8\pi G}{c^4} \cdot M \tag{C4}$$

Substituting Eq. (C4) into Eq. (C1), we obtain

$$R^{ij} = -\frac{8\pi G}{c^4} \cdot M^{ij} + \frac{1}{2} \cdot g^{ij} R = -\frac{8\pi G}{c^4} \cdot \left(M^{ij} - \frac{1}{2} \cdot g^{ij} M\right). \tag{C5}$$

Using antisymmetric tensor f_{ij} which denotes the magnitude of electromagnetic field, the electromagnetic energy tensor M^{ij} is represented as follows:

$$M^{ij} = -\frac{1}{\mu_0} \cdot \left(f^{ip} f_\rho^j - \frac{1}{4} \cdot g^{ij} f^{\alpha\beta} f_{\alpha\beta} \right), \quad f^{ip} = g^{i\alpha} g^{\rho\beta} f_{\alpha\beta} \ . \tag{C6}$$

Therefore, for M, we have

$$\begin{aligned} M = M_i^i = g_{ij} M^{ij} &= -\frac{1}{\mu_0} \cdot \left(g_{ij} f^{ip} f_\rho^j - \frac{1}{4} \cdot g_{ij} g^{ij} f^{\alpha\beta} f_{\alpha\beta} \right) \\ &= -\frac{1}{\mu_0} \cdot \left(f^{ip} f_{ip} - \frac{1}{4} \cdot 4 f^{\alpha\beta} f_{\alpha\beta} \right) = -\frac{1}{\mu_0} \cdot \left(f^{ip} f_{ip} - f^{ip} f_{ip} \right) = 0 \end{aligned} \tag{C7}$$

Accordingly, substituting $M = 0$ into Eq. (C5), we get

$$R^{ij} = -\frac{8\pi G}{c^4} \cdot M^{ij} \ . \tag{C8}$$

Although Ricci tensor R^{ij} has 10 independent components, the major component is the case of $i = j = 0$, i.e., R^{00}. Therefore, Eq. (C.8) becomes

$$R^{00} = -\frac{8\pi G}{c^4} \cdot M^{00} \ . \tag{C9}$$

On the other hand, 6 components of antisymmetric tensor $f_{ij} = -f_{ji}$ are given by electric field E and magnetic field B from the relation to Maxwell's field equations

$$f_{10} = -f_{01} = \frac{1}{c} \cdot E_x, f_{20} = -f_{02} = \frac{1}{c} \cdot E_y, f_{30} = -f_{03} = \frac{1}{c} E_z$$
$$f_{12} = -f_{21} = B_z, f_{23} = -f_{32} = B_x, f_{31} = -f_{13} = B_y \quad \text{(C10)}$$
$$f_{00} = f_{11} = f_{22} = f_{33} = 0$$

Substituting Eq. (C10) into Eq. (C6), we have

$$M^{00} = -\left(\frac{1}{2} \cdot \varepsilon_0 E^2 + \frac{1}{2\mu_0} \cdot B^2\right) \approx -\frac{1}{2\mu_0} \cdot B^2. \quad \text{(C11)}$$

Finally, from Eqs. (C9) and (C11), we get

$$R^{00} = \frac{4\pi G}{\mu_0 c^4} \cdot B^2 = 8.2 \times 10^{-38} \cdot B^2 \quad (B \text{ in Tesla}), \quad \text{(C12)}$$

where we let $\mu_0 = 4\pi \times 10^{-7} (H/m)$, $\varepsilon_0 = 1/(36\pi) \times 10^{-9} (F/m)$, $c = 3 \times 10^8 (m/s)$, $G = 6.672 \times 10^{-11} (N \cdot m^2 / kg^2)$, B is a magnetic field in Tesla and R^{00} is a major component of spatial curvature $(1/m^2)$.

The relationship between curvature and magnetic field was derived by Minami and introduced it in 16th International Symposium on Space Technology and Science (1988) [1].

Eq. (C12) is derived from general method.

On the other hand, Levi-Civita also investigated the gravitational field produced by a homogeneous electric or magnetic field, which was expressed by Pauli [2]. If x^3 is taken in the direction of a magnetic field of intensity F (Gauss unit), the square of the line element is of the form;

$$ds^2 = (dx^1)^2 + (dx^2)^2 + (dx^3)^2 + \frac{(x^1 dx^1 + x^2 dx^2)^2}{a^2 - r^2}$$
$$- \left[c_1 \exp(x^3/a) + c_2 \exp(-x^3/a)\right]^2 (dx^4)^2 \quad \text{(C13)}$$

where $r = \sqrt{(x^1)^2 + (x^2)^2}$, c_1 and c_2 are constants, $a = \dfrac{c^2}{\sqrt{kF}}$, k is Newtonian gravitational constant(G), and $x^1...x^4$ are Cartesian coordinates ($x^1...x^3$ = space, $x^4 = ct$) with orthographic projection.

The space is cylindrically symmetric about the direction of the field, and on each plane perpendicular to the field direction the same geometry holds as in Euclidean space on a sphere of radius a, that is, the radius of curvature a is given by

$$a = \frac{c^2}{\sqrt{kF}} \quad . \tag{C14}$$

Since the relation of between magnetic field B in SI units and magnetic field F in CGS Gauss units are described as follows: $B\sqrt{\dfrac{4\pi}{\mu_0}} \Leftrightarrow F$, then the radius of curvature "a" in Eq. (C14) is expressed in SI units as the following (changing symbol, $k \rightarrow G, F \rightarrow B$):

$$a = \frac{c^2}{\sqrt{GF}} = \frac{c^2}{\sqrt{G} \cdot B \sqrt{\dfrac{4\pi}{\mu_0}}} \approx (3.484 \times 10^{18} \frac{1}{B} \ meters) \quad . \tag{C15}$$

While, scalar curvature is represented by

$$R^{00} \approx R = \frac{1}{a^2} = \frac{GB^2 \dfrac{4\pi}{\mu_0}}{c^4} = \frac{4\pi G}{\mu_0 c^4} B^2 \ , \tag{C16}$$

which coincides with (C12).

References

[1] Minami, Y., "Space Strain Propulsion System", *16th International Symposium on Space Technology and Science (16th ISTS)*, Vol.1, 1988: 125-136.

[2] Pauli, W. *Theory of Relativity*, Dover Publications, Inc., New York, 1981.

APPENDIX D: PROPERTIES OF HYPER-SPACE

Let us put x^1, x^2, x^3 for x, y, z and x^0 for *ct*. In Minkowski space, the distance(s) are given by

$$S_{RS}^2 = \eta_{ij}x^i x^j = -(x^0)^2 + (x^1)^2 + (x^2)^2 + (x^3)^2 = -(ct)^2 + x^2 + y^2 + z^2 \quad (D1)$$

where η_{ij} is Minkowski metric, and *c* is the speed of light.

Eq. (D1) indicates the properties of the actual physical space limited by Special Relativity.

From Eq. (D1), Minkowski metric, i.e., real-space metric is shown as follows:

$$\eta_{ij(RS)} = \begin{bmatrix} -1 & 0 & 0 & 0 \\ 0 & +1 & 0 & 0 \\ 0 & 0 & +1 & 0 \\ 0 & 0 & 0 & +1 \end{bmatrix},$$

$$\eta_{00} = -1, \ \eta_{11} = \eta_{22} = \eta_{33} = +1, \text{ and other } \eta_{ij} = 0. \quad (D2)$$

The properties of space are determined by the metric tensor, which defines the distance between two points.

Appendices

Here, as a hypothesis, we demand an invariant distance for the time component of Minkowski metric reversal, i.e., $\eta_{ij} \to \eta_{ij}$, $\eta_{00} \to -\eta_{00}$.

This hypothesis gives the minimum distance between two kinds of space-time, that is, the properties of Hyper-Space are close to that of Minkowski space.

Namely, regarding the following equation;

$$(x^0)^2 \to -(x^0)^2 \ [= (ct)^2 \to -(ct)^2], \text{ i.e., } \eta_{00} = -1 \to \eta_{00} = +1,$$

we require the following equation.

$$S_{HS}^{\ 2} = S_{RS}^{\ 2}, \tag{D3}$$

where suf."HS" denotes Hyper-Space and suf."RS" denotes Real-Space.

From the above hypothesis, the metric of Hyper-Space becomes as follows:

$$\eta_{ij(HS)} = \begin{bmatrix} +1 & 0 & 0 & 0 \\ 0 & +1 & 0 & 0 \\ 0 & 0 & +1 & 0 \\ 0 & 0 & 0 & +1 \end{bmatrix},$$

$\eta_{00} = +1$, $\eta_{11} = \eta_{22} = \eta_{33} = +1$, and other $\eta_{ij} = 0$. (D4)

Therefore, in Hyper-Space, the distance is given by

$$S_{HS}^{\ 2} = \eta_{ij} x^i x^j = (x^0)^2 + (x^1)^2 + (x^2)^2 + (x^3)^2 = (ct)^2 + x^2 + y^2 + z^2. \tag{D5}$$

Accordingly, Hyper-Space shows the properties of Euclidean space. Therefore, the imaginary time ($x^0 = ict$; $i^2 = -1$) as the component of time coordinate is required as a necessary result in Hyper-Space. Because, by substituting the imaginary time "it" into Eq. (D5), we get Eq. (D1), so that the invariance of distance is satisfied. The time "t" in Real-Space is changed to "it" in Hyper-Space. Here, "i" denotes the imaginary unit ($\sqrt{-1}$). However, the components of space coordinates (x,y,z) are the same real numbers as Real-Space.

INDEX

A

Acceleration, 6, 10, 12, 18, 37, 59, 64, 72, 73, 74, 75, 86, 87, 88, 89, 90, 91, 92, 94, 95, 97, 99, 102, 103, 104, 112, 116, 118, 125, 127, 128, 129, 130, 134, 143, 163, 164, 172, 173, 174, 175, 179, 181, 182, 185, 192, 208, 223, 228, 230, 231, 232, 233, 234
  equation, 112, 116
  field, 18, 72, 74, 75, 86, 87, 89, 90, 92, 97, 103, 118
  mechanism, 125, 130
accretion disk, 123, 124, 125, 127, 128, 129, 141, 142, 147, 148, 149, 150, 151
active galactic nucleus, 123
advanced propulsion system, 158, 210
Alqubierre Warp Drive, 182
Annihilation, 4, 10, 15, 22, 38, 51, 144, 145, 149, 150
anti-particle, 10
anti-symmetric curvature, 92
area element, 84, 241
astronomy, ix, 127, 193
astrophysical jet, 123, 124, 125, 127, 141, 151
astrophysical phenomena, 123, 143
astrophysics, ix, 141, 143, 150
avalanche-ionization processes, 141

B

base vector, 80
beamed energy, 7, 64, 185
black holes, 123, 124, 147, 149
black vortex, 141, 142
body force, 10, 72, 82, 84, 85, 99, 119
breaking of magnetic field lines, 124

C

Casimir effect, 8, 20, 22, 36
Casimir force, 8, 20
causality principle, 206
charged particle, 103, 124, 142, 150, 151
coherent state, 22, 23, 24, 32
collimation mechanism, 125, 127
collision, 145, 149

computational fluid dynamics (CFD), 51, 54, 185
conic surface, 172, 173, 174, 180
conservation law, 5, 9, 95, 96, 97, 181
continuity equation, 97, 98
continuum mechanics, 3, 4, 15, 16, 71, 73, 80, 81, 82, 83, 84, 85, 241
coordinate system, 80, 83, 87, 88, 133, 200, 227
cosmic time, 112, 113
cosmological constant, 72, 100, 101, 102, 106, 110, 111, 112, 113, 114, 116, 234, 235
cosmological term, 72, 99, 100, 102, 106, 117, 234
cosmology, viii, 18, 72, 73, 99, 104, 108, 112, 120, 122, 212, 234
covariant derivative, 80, 83, 238
covariant differentiation, 80, 82, 83, 84
creation, 4, 10, 15, 22, 51
  operator, 22
Curvature, 4, 5, 10, 14, 15, 17, 33, 39, 69, 71, 72, 73, 74, 75, 76, 77, 78, 85, 86, 87, 88, 89, 90, 94, 101, 106, 108, 110, 133, 134, 135, 141, 172, 225, 226, 227, 228, 230, 238, 239, 244, 245
  of space, 4, 15, 18, 69, 71, 72, 73, 75, 78, 85, 86, 87, 101, 225, 227, 230, 239
curved space, 4, 15, 17, 72, 74, 76, 78, 83, 86, 87, 88, 89, 90, 91, 92, 94, 95, 96, 118, 171, 225, 226, 227, 228, 230
  region, 72, 86, 87, 88, 90, 91, 94, 95, 96, 118, 228
curved surface, 17, 74, 76, 77, 86, 226

D

de Sitter, 16, 71, 72, 73, 85, 90, 100, 101, 102, 108, 111, 120, 152, 210, 213, 234
  line element, 101, 234
  solution, 72, 73, 90, 100, 102, 152, 234

de-Broglie waves, 161
deformation, 16, 18, 71, 79, 80, 82, 85, 97, 236, 239
deformed space, 79
disordered phase, 20, 105
displacement vector, 80, 239

E

early universe, 20, 105, 106
Earth, vii, viii, xi, xiii, 6, 37, 51, 52, 61, 64, 65, 67, 96, 97, 107, 121, 122, 130, 152, 153, 157, 158, 159, 163, 164, 167, 173, 174, 181, 185, 190, 191, 193, 198, 208, 209, 224, 228
elastic body, 4, 5, 15, 16, 18, 71, 85, 96
elastic field, 16, 71, 85, 97
elastic force, 18, 19, 82, 84
elastic modulus, 82
elastic solid phase, 105
electric field, 8, 20, 24, 25, 28, 90, 144, 170, 243
electric propulsion, 7, 12
electromagnetic energy, 40, 51, 81, 82, 185, 227, 242, 243
electromagnetic field, 4, 15, 24, 103, 243
Electron, 63, 124, 128, 142, 149, 150, 151, 160
  -positron pairs, 124, 128, 142, 149, 150, 151
  -positron production, 142
electroweak interaction, 72, 109
elementary particle physics, 104
EM field, 64, 165, 167, 168, 183, 186
EM wave energy, 168
empty space, 39, 40, 44, 49, 101, 235
energy density, 4, 8, 9, 15, 19, 21, 27, 28, 29, 32, 40, 97, 98, 104, 106, 110, 111, 112, 113, 114, 116, 233
energy momentum tensor, 77, 100, 101, 102, 110, 113, 117, 227, 235

Index

energy-conversion efficiency, 146
engine, 11, 13, 14, 20, 63, 91, 96, 98, 99, 102, 124, 130, 131, 134, 135, 141, 152, 186, 187
Entropy, 5, 10, 14, 16, 18, 19, 20, 105, 171, 174
  elasticity, 18
equilibrium condition, 75, 82, 84
equilibrium state, 13, 90, 91
Euclidean metric, 217
event horizon, 127
excitation of space, 20, 105, 106, 152
Expanded Maxwell Equations, 167
expanding universe, 16, 71, 85, 108, 111, 120, 225
expanding vacuum space, 117
Expansion, vii, 4, 15, 16, 54, 55, 71, 72, 73, 85, 100, 105, 108, 109, 116, 118, 119, 190
  of space, 118, 119
expectation value, 27, 29, 40, 41, 103, 106
external field, 1, 3, 69
extrasolar planet, 157

F

Faster-Than-Light (FTL), viii, xi, 54, 66, 161, 162, 163, 166, 167, 171, 174, 175, 177, 179, 181, 183, 186, 194, 196, 224
Feynman Kernel, 218
field power and propulsion, xi, 63, 187, 188, 224
field propulsion, viii, ix, xi, 1, 2, 3, 4, 5, 7, 9, 10, 11, 12, 13, 14, 15, 49, 59, 61, 63, 65, 67, 69, 70, 73, 143, 144, 186, 190, 194, 223, 224
  system, viii, ix, 1, 2, 5, 7, 9, 10, 11, 13, 14, 69, 70, 143, 186, 223
field-propelled flight, 185
fine structure of space-time, 5, 16

fine-grained structure technology, 213, 214, 215, 216
fine-structure of space, 85
finite interval of time, 13, 90
flat space, 11, 78, 83, 87, 90, 91, 92, 104, 117, 226, 228, 230
fluid equation, 112, 113, 114
four velocity, 87
fracture point, 210
Friedmann, 16, 71, 85, 108, 111, 112, 113, 116, 120
frozen-in magnetic field, 136
fusion reactor, 59, 60, 61, 62

G

galaxy exploration, 152, 221
gas effective exhaust speed, 5, 6
Gaussian curvature, 74, 76
General Relativity, 1, 2, 3, 4, 12, 15, 54, 69, 71, 73, 85, 87, 108, 111, 121, 122, 159, 182, 194, 195, 225, 226, 227, 228, 230, 231
geodesic equation, 87, 226
geodesic line, 87, 226
geometrical structural deformation, 81, 82
gluon, 44, 45, 49, 51
gravitation, 4, 8, 10, 15, 117, 121, 225
gravitational action, 82, 85
gravitational constant, 77, 100, 110, 211, 227, 232, 245
gravitational energy, 124, 146, 147, 148, 149
gravitational field equation, 71, 73, 77, 86, 89, 100, 110, 117, 227, 229, 231, 235
gravitational power plant, 124

H

habitable zone, 157, 158, 209
helical magnetic field, 125, 129

Higgs field, 48, 102, 103, 115
Higgs quantum vacuum, 48, 49, 51
high energy radiation, 123
high temperature plasma, 149
Hubble constant, 112
Hubble parameter, 112
Hyperspace, vi, ix, 160, 165, 166, 167, 193, 195, 197, 198, 199, 200, 201, 202, 203, 204, 205, 206, 207, 208, 209, 210, 213, 214, 215, 216, 217, 218, 220, 224, 246, 247
    navigation theory, 160, 195, 197, 210

I

imaginary time, 195, 198, 199, 203, 205, 206, 207, 208, 210, 217, 218, 247
    hole, 198
inertia, 8, 10, 51, 53, 64, 165, 167, 228
inertial force, 10, 72, 99, 119, 137
infinitesimal distance, 79, 80, 81, 228, 229, 236
inflationary cosmological model, 16, 71, 85, 108, 116, 120
inflationary phase, 116
inhomogeneous field, 8, 20
interaction of space-time, 1, 3, 8, 12, 71
intergalactic exploration, viii, 160, 194, 223
interplanetary travel, 1
interstellar magnetic field lines, 127, 137
interstellar travel, 1, 57, 108, 143, 144, 157, 159, 160, 194, 195, 197, 198, 204, 205, 207, 209, 210

J

jumping, viii, 160, 186, 194

K

Kerr solution, 73

L

Lagrangian, 219
laser beam, 25, 31, 33, 136, 137
laser propulsion, 7, 12
light radiation pressure, 125
light speed, 37, 165
line element, 79, 87, 236, 238, 239, 240, 244
line stress, 17, 74, 76, 77, 86
Liquid Metal MHD Power Generation System Using Antiproton Annihilation Reactor, 143, 144
locally-expanded space, 117
Lorentz transformation, 102, 199, 200
Lorentz-Fitz Gerald contraction factor, 200
lower-dimensional spacetime, 161

M

magnetic centrifugal force, 125, 128
magnetic field, 59, 60, 72, 86, 90, 91, 92, 94, 99, 104, 107, 124, 125, 126, 127, 128, 129, 130, 133, 134, 135, 136, 141, 150, 151, 243, 244, 245
    line, 124, 125, 126, 127, 128, 129, 130, 150, 151
    pressure, 125
    reconnection, 124, 141, 150, 151
magnetic flux compression technology, 136
magnetic flux line breaking, 141
magnetic flux reconnection, 142
magnetic force, 125
magnetic pressure, 59, 125, 128, 129
magnetic tension, 125

magneto-hydrodynamic (MHD), 61, 62, 63, 127, 128, 144, 145
major component of spatial curvature, 74, 85, 86, 88, 244
Many-Particle Systems, 207, 210
mass energy, 81, 82, 146, 227
mass expulsion, 1, 8, 12
mass ratio, 5, 6, 8
matter-emitting propulsion, 59
mechanical property of space-time, 1, 3
metastable false vacuum, 106
metric, 10, 14, 49, 54, 67, 76, 77, 80, 81, 82, 86, 88, 89, 102, 108, 109, 159, 167, 175, 182, 183, 194, 198, 223, 225, 226, 227, 228, 229, 230, 231, 242, 246, 247
metric tensor, 76, 77, 80, 81, 82, 86, 88, 89, 198, 225, 226, 227, 228, 229, 230, 242, 246
Minkowski metric, 199, 217, 228, 246, 247
Minkowski space, 161, 198, 200, 201, 246, 247
momentum conservation law, 5, 6, 12, 95, 96, 97, 98
momentum thrust, 1, 2, 3, 5, 7, 8, 9, 12, 13, 14, 69, 95, 96, 194

N

natural unit, 104
navigation system, 158, 205, 210
negative pressure, 56, 114
neutron stars, 123, 147, 149
normal stress, 17, 73, 74, 76, 77, 86
nuclear fusion, 48, 50, 59, 60, 146, 147, 187
nuclear propulsion, 7, 12, 30

O

orbital speed, 6, 61, 63
ordered phase, 20, 105

P

parallel space, 206, 207, 210, 217
parametric amplifier, 31
partial derivative, 80, 83
Path Integrals, 207, 217
payload, 7
phase transition, 72, 105, 106, 109
Photon, 3, 29, 38
  rocket, 38
physical structure of space, 1
pinch, 125, 127, 129
Planck energy, 211, 216
Planck length, 39, 194, 211, 212, 213, 216
Planck mass, 211, 212, 213, 216
Planck time, 211, 212, 216
plasma gas, 123, 127, 130, 147, 149
polarization modulation, 168, 169, 170
polarizing of quantum vacuum, 165
polarizing zero-point vacuum, 167
polymer chains, 5, 16, 18
positron-electron plasma, 127
potential barrier, 206, 207, 210, 211, 212, 213, 214, 216, 217, 218
potential gradient, 2, 9, 10, 13, 14, 70
power source, 34, 90, 91, 98, 99, 124, 141, 142, 143, 152, 159, 217
pressure field, 17, 71, 73, 74, 75, 77, 86
pressure gradient, 2, 9, 10, 12, 13, 14, 57, 70, 185
pressure thrust, 1, 2, 3, 8, 9, 12, 13, 69, 71, 95, 96, 194
principal curvature, 17, 74, 76, 77, 86
principal radii, 17, 74, 77
Propellant, vii, viii, xi, 6, 7, 14, 51, 59, 60, 63, 65, 67, 188, 190, 223, 224
  -less power and propulsion, 51, 188, 190
  -less propulsion, 14, 59
proper time, 87
propulsion mechanism, viii, 90, 91, 95, 96

propulsion principle, 2, 3, 4, 5, 6, 7, 9, 11, 12, 13, 14, 15, 30, 69, 70, 71, 73, 91, 96, 108, 117, 118, 120, 122, 124, 137, 142, 143, 159

propulsion theory, viii, 1, 3, 11, 69, 71, 108, 159, 160, 194, 197, 207

propulsive force, 1, 2, 3, 4, 8, 12, 15, 71, 96, 223

proton, 43, 44, 45, 49, 50, 144

pulse propulsion system, 10, 92

Q

quadratic surface, 74, 75, 76

Quantum Chromo Dynamic (QCD), 43, 44, 45, 48, 49, 50, 51, 57

Quantum Cosmology, 12

quantum electrodynamics (QED), 4, 15, 48, 51

Quantum Field Theory, 1, 2, 3, 4, 5, 12, 15, 16, 18, 22

Quantum Interstellar Ramjet, 43, 57

quantum noise, 8, 21, 22, 24

quantum optics, 4, 15, 22

quantum state, 5, 16, 29

quantum tunneling effect, 212, 213, 214, 215

quark, 45, 48, 49

quasi-black hole, 141

quasi-light speed, 11

R

ramjet, 38, 39, 40, 41, 42, 47, 48, 51

rapid expansion of space, 72, 73, 109

reaction mass, 9

reaction thrust, 8, 12, 14, 95

Real-Space, 198, 199, 200, 201, 202, 203, 204, 205, 206, 207, 210, 214, 215, 217, 218, 220, 247, 248

Reissner-Nordstrom solution, 73

relativistic jets, 127

Ricci tensor, 77, 78, 88, 100, 110, 227, 229, 230, 243

Riemann curvature tensor, 74, 78, 83, 85, 90, 228, 229, 230, 238, 241

Riemann space, 74, 78, 83, 87, 227, 230

Riemannian connection coefficient, 78, 87, 89, 110, 230

Riemannian geometry, 71, 78, 85, 225, 227, 230

Robertson-Walker metric, 73, 109, 110

rocket, vii, 5, 6, 7, 37, 38, 39, 43, 48, 60, 61, 63, 96, 97, 186, 187

rotating vortex, 141

rotation tensor, 83, 241, 242

Rubber, 15, 16, 18, 19, 71, 85, 105, 129

elasticity, 18, 19

S

scalar curvature, 72, 77, 78, 100, 101, 110, 227, 229, 245

scalar field, 102, 103, 106, 115

scale factor, 109, 110, 112, 117

Schrödinger equation, 220

Schwarzschild solution, 72, 73, 152, 232, 233

seed magnetic field, 91, 136, 137

shock wave, 125

Slower-Than-Light (STL), 161, 162, 163, 166, 167, 171, 174, 175, 177, 181, 182, 184, 195

Space, v, vii, viii, ix, xi, xiii, xiv, 1, 2, 3, 4, 5, 7, 9, 10, 12, 13, 15, 16, 17, 18, 19, 20, 23, 32, 34, 35, 36, 37, 39, 40, 43, 44, 45, 46, 47, 49, 51, 52, 57, 59, 62, 64, 65, 67, 69, 70, 71, 72, 73, 74, 75, 78, 79, 81, 82, 83, 85, 86, 87, 89, 90, 91, 92, 94, 95, 96, 97, 98, 99, 100, 101, 102, 104, 105, 106, 107, 108, 109, 110, 113, 114, 115, 117, 119, 120, 121, 122, 124, 130, 132, 133,

Index

134, 135, 141, 142, 143, 145, 150, 151, 152, 153, 158, 159, 160, 162, 163, 164, 165, 166, 167, 171, 172, 173, 175, 176, 177, 178, 181, 182, 183, 186, 187, 188, 190, 192, 194, 195, 197, 198, 199, 200, 201, 202, 203, 205, 206, 207, 208, 209, 210, 211, 213, 214, 215, 217, 218, 220, 221, 223, 224, 225, 226, 227, 228, 229, 230, 231, 233, 236, 239, 242, 244, 245, 246, 247
- drive, viii, 10, 69, 70, 71, 72, 73, 85, 89, 90, 91, 95, 96, 98, 99, 100, 106, 108, 109, 117, 119, 122, 124, 133, 142, 143, 152, 225
- drive propulsion, viii, 69, 70, 71, 72, 73, 85, 89, 90, 91, 95, 96, 98, 99, 100, 106, 108, 109, 117, 119, 122, 124, 133, 142, 143, 152, 225
- drive propulsion device, 133
- drive propulsion system, 69, 70, 71, 72, 89, 90, 96, 99, 100, 106, 108, 124, 152
- flight, xiii, 45, 62, 65, 159, 187, 194
- navigation theory, 158, 210
- propulsion theory, 158, 210
- strain, 10, 71, 79, 85

spacecraft, 6, 59, 61, 137, 141, 142, 155
spaceship, 1, 2, 3, 6, 8, 12, 13, 14, 20, 34, 69, 70, 71, 72, 90, 91, 92, 94, 95, 96, 97, 98, 99, 103, 117, 118, 119, 120, 134, 135, 143, 159
Space-Time, 122, 160, 167, 171, 177, 182, 220
- fabric, 4, 16, 18
- structure, 1, 3, 8, 12
- -warping, 164, 182

spatial curvature, 17, 18, 74, 86, 87, 89, 90, 91, 94, 98, 110, 134, 152, 228
Special Relativity, 37, 54, 85, 87, 161, 172, 176, 180, 193, 195, 197, 198, 200, 201, 204, 205, 210, 225, 226, 228, 246
specially-conditioned EM field, 171, 175
specific impulse, 6
speed of light, 10, 11, 13, 39, 88, 95, 100, 119, 123, 125, 127, 157, 159, 193, 197, 199, 205, 207, 208, 227, 246
spherical coordinate, 89, 231
spherical space, 17, 74, 77
spontaneous symmetry breaking, 106
squeeze operator, 22, 24
squeeze parameter, 22, 24, 27, 28, 29, 30, 33
squeezed light, 4, 9, 15, 21, 22, 25, 27, 28, 29, 30, 31, 32, 33, 34
squeezed vacuum, 4, 9, 15, 21, 22, 23, 24, 25, 27, 28, 29, 31, 32
- energy density, 27, 29
- state, 4, 9, 15, 21, 22, 23, 24, 25, 27, 28, 31, 32

squeezing, 23, 24, 28, 33, 36
starship, 9, 10, 12, 30, 39, 160, 193, 194, 197, 200, 201, 202, 203, 204, 205, 206, 207, 208, 210, 211, 212, 213, 214, 215, 216, 217, 218, 219, 220
stellar system, 155, 156, 159, 193, 197
strain field, 81, 82
strain propagation velocity, 91
strain rate, 13, 90, 91, 95, 207
strain tensor, 81, 82, 83, 84, 226
stress field, 82
stress tensor, 29, 82, 84, 85
stress-energy tensor, 72
stress-free acceleration, 181
strings, 4, 16, 18, 20
strong magnetic field, 71, 72, 90, 91, 123, 125, 134, 136, 137, 141, 142, 151, 152
SU(2), 48, 167, 168, 169
substantial physical structure, 1, 3, 4, 5, 15, 73, 109
super-Earth, 157
superstring theory, 4, 16, 18, 194
surface force, 17, 73, 74, 75, 77, 82, 85, 86
symmetric curvature, 92
synchrotron radiation, 141, 149, 150

Index

T

Tachyons, 161, 162, 163, 191, 194
Tardyons, 161, 162, 163
Tensor, 78, 80, 81, 82, 83, 88, 100, 102, 167, 225, 226, 227, 228, 229, 230, 238, 241, 242, 243
  equation, 80, 83
thermal gas pressure, 125
thrust, viii, xi, 1, 2, 3, 4, 7, 8, 9, 10, 12, 14, 15, 20, 21, 33, 34, 41, 43, 57, 60, 61, 63, 69, 72, 90, 91, 92, 94, 95, 96, 97, 99, 108, 118, 119, 135, 142, 223
time gap, 159, 204, 205
time paradox, 159, 195, 204, 205
transmissivity, 213, 214
twisted magnetic field, 129

U

U(1), 168
UK Patent, 130, 132, 137
ultra-high temperature, 20
uncertainty principle, 215, 216
Urashima effect, 159

V

Vacuum, viii, 1, 2, 3, 4, 7, 8, 9, 10, 12, 13, 15, 16, 18, 19, 20, 21, 22, 23, 24, 25, 26, 27, 29, 30, 31, 32, 33, 35, 36, 39, 40, 41, 43, 44, 45, 46, 47, 48, 49, 50, 51, 52, 53, 54, 55, 56, 57, 64, 67, 70, 71, 72, 73, 85, 99, 100, 101, 102, 103, 104, 105, 106, 108, 109, 113, 114, 115, 117, 118, 119, 145, 150, 151, 165, 170, 175, 180, 184, 185, 186, 192, 234

energy density, 27, 30, 31, 102
expectation value, 102, 103, 104, 106
field, 2, 9, 12, 13, 45, 48, 51, 70, 114, 115, 118, 119
perturbation, 4, 8, 15, 21, 27, 29, 30, 31, 33
potential, 102, 103, 106, 114
state, 4, 8, 9, 15, 21, 22, 23, 24, 25, 27, 32
vector parallel displacement, 83, 230
velocity of light, 77, 86, 87, 88, 89, 90, 91, 92, 99, 110, 159, 202, 204, 210, 211, 213, 214, 216, 217, 230
virtual particle pairs, 4, 15
visco-elastic liquid phase, 105
viscosity, 147, 148, 149

W

wave function, 214, 217, 220
wave packet, 162, 163
Weinberg-Salam model, 72, 109
white dwarfs, 123
WKB approximation, 213, 214
wormhole, 159, 194, 197, 198, 208

Z

zero point oscillator, 20, 21, 34
zero point radiation pressure, 20
zero-point energy, 8, 20, 22, 40, 41, 48, 51, 185
Zero-Point Field (ZPF), 8, 10, 12, 21, 24, 27, 141, 142
zero-point fluctuation, xiii, 4, 5, 15, 24, 40
zero-point radiation pressure, 10, 14, 21, 30